The Oil Business in
Latin America

The Oil Business in
Latin America
‖ *The Early Years* ‖

EDITED BY JOHN D. WIRTH

BeardBooks
Washington, D.C.

Library of Congress Cataloging-in-Publication Data

Latin American oil companies and the politics of energy.
   The oil business in Latin America : the early years / edited by John D. Wirth.
         p. cm.
   Originally published: Latin American oil companies and the politics of energy.
Lincoln, Neb.: University of Nebraska Press, 1985. (Latin American studies series)
      ISBN 1-58798-103-3 (pbk.)
      1. Petroleum industry and trade--Latin America--Case studies. 2. Petroleum industry
and trade--Government ownership--Latin America--Case studies. I. Wirth, John D. II.
Title.

HD9574.L3 L38 2001
338.7'6223382'098--dc21
[B]   2001035698

Printed in the United States of America

Contents

ILLUSTRATIONS

Preface

The writing and editing of these essays was greatly facilitated by mutual criticism among the authors, who passed chapter drafts back and forth during the four-year life of this collaborative project. Sustained interaction began when Edwin Lieuwen, Carl Solberg, and John Wirth prepared papers for a panel at the Los Angeles meetings of the American Historical Association in 1981. Alfred Saulniers commented on the papers (along with Peter S. Smith) and was promptly commissioned to write our conclusion. Michael Meyer chaired the session and promptly signed us up for his Latin American Studies Series at the University of Nebraska Press, of which this is the third volume. In Los Angeles, Esperanza Durán joined our group and planned her chapter on Mexico, as did Jonathan Brown, whose research on Standard Oil of New Jersey appears in the second chapter of this book. The editor wishes to thank Iris Salinas for preparing the production and consumption graphs in the Introduction.

John D. Wirth

Introduction

When Argentina established the world's first state oil company in 1922, it began a trend which fifty years later was almost the norm worldwide: governments own, control, manage, and exploit national resources for national ends, in name of the common good. Everywhere in Latin America this fundamental tenet of economic nationalism is in effect. The state companies are also considered a strategic arm of public policy, a resource for nation building. Once dominant in all phases of the business, the large multinational petroleum companies are now subordinate to the nation state where they operate. In this first and most celebrated of confrontations between the state and multinational enterprise, the state won.[1] Governments can choose from a variety of smaller oil companies, contractors, and suppliers. Private national companies, when they are active in the Latin American industry at all, are at best a supplement to the state oil companies, not a substitute for them or an alternative.

Yet the economic and political consequences of this statist trend in Latin America are still being worked out after half a century of experience and practice with the state enterprise. The economics of this highly capital-intensive industry have compelled the tempering of ideology with pragmatism. With the partial exceptions of Mexico and Venezuela, no Latin American country could reasonably expect to finance exploration and development itself. Service contracts and various forms of production agreements (but not old-style concessions) came in with high energy prices in the 1970s. And today the four countries featured in this book, which collectively account for 85 percent of Latin America's crude oil production and three-quarters of its petroleum consumption, are all deeply in debt. Given the large public debt, it is clear that private capital does have a role to play in a continent whose petroleum potential is still largely unknown. Seventy years after the first major fields were developed, in Mexico, much remains to be done.

Oil nationalism is part of the very definition of nationhood in Latin America. For many, the state oil companies are a source of national pride, virtually a birthright, and the strategic core of a distinctive model of development. Yet the belief that sovereignty over natural resources would assure political and economic autonomy has been overtaken by events. The volatile world oil market, coupled with policies of debt-led growth, rendered obsolete earlier conceptions of national self-sufficiency. In most countries today the debate centers on what mix of the nationalist and the free market approach is the most adequate for each nation. It so happens that this is an old debate in Latin America, well covered in our four case studies. What seems to be new is less emphasis on the risks of cooperation, at a time when high oil prices have played havoc with the region's economies. Today, the very existence of state companies helps assure that foreign enterprise cannot take undue advantage of the nation. In turn, foreign companies still have the advantages of global integration, with the financial and technical resources this provides.[2] In an increasingly competitive and unforgiving world economy, devising the appropriate industrial policy rather than controlling and developing natural resources is the more relevant strategy to protect national independence.

Regional integration, another theme of the early nationalists, has not yet developed much beyond the setting up of organizations. Lack of capital as well as market forces are both propelling integration, however, and there has been some success in the regional acquisition and management of technology. The latest proposal is Petrolatín, which would become a multinational company based on a pooling of managerial, technical, and marketing resources of Latin America's three largest public enterprises: Brazil's Petrobrás, Venezuela's PDVSA and Mexico's Pemex. Petrolatín would benefit from a long history of shared ideas, perceptions, and aspirations. In fact, one theme in the chapters below is the point-counterpoint of national experiences and regional thinking. Put another way, each national company developed differently but within a regional context. Thus, as Latin American energy markets begin to be integrated across national borders today, it is well to remember that an agenda for regional integration is not new, even if practical frameworks to accomplish this are still illusive.

Looking back from the complex mix of dependencies and interdependencies that drives Latin America today, we think that the historical dynamics of this industry should be reexamined. Much has been written on the Latin American petroleum industry, including previous studies by Carl Solberg, Edwin Lieuwen, John Wirth, and Esperanza Durán. Here we draw upon new evidence to assess the ways and circumstances in which the state companies were founded and to evaluate the consequences: in short, to explore their *historicity*. Also included is a new perspective on the early Latin American operations of Standard Oil of New Jersey.

Our approach is conceptual and problem oriented, rather than narrative. In the chapters on Standard Oil, Argentina, Brazil, Mexico, and Venezuela each author moves from an assessment of current conditions back to the critical formative years for each of the companies under review. The chronology depends on how each author defines the problem. For example, to Jonathan Brown, the formative years for Standard Oil were before 1930, whereas for Lieuwen the elaboration of Venezuela's PDVSA is very much a postwar phenomenon. Analysis of the Argentine company YPF by Solberg, of the pre-Pemex years by Durán, and of the formative period before Petrobrás by Wirth will focus primarily on the 1930s. In conclusion, Alfred Saulniers uses a public policy perspective to evaluate how well the state companies have fulfilled their goals.

To introduce these essays on origins, this chapter will discuss briefly the industry in each country and then move to define the state enterprise and its functions before concluding with a short history of petroleum integration.

Overview of the Industry

The changing importance of Latin America in the world petroleum markets is well covered by George Philip in his recent comprehensive survey. Suffice it to say here that in the early twentieth century first Mexico and then Venezuela became major producers, but then the region was relegated to secondary status when the Middle Eastern fields came into production by the end of World War II. The rise of Mexico as the world's fourth-largest producer in the late 1970s once again focused attention on Latin America's actual and potential role in world markets. Even more important, the postwar industrialization of this region, however uneven, created a large internal demand for petroleum products. Even Venezuela, the classic example of a petroleum exporter in the hemisphere, devoted more of its resources to serving the domestic market. The four state companies under review became giant industrial enterprises in their own right, each in its own way essential to maintaining and expanding economic development. The graphs show the growing importance of the internal market, as well as the relative weight of imports and exports in the petroleum history of each country.

Historically, the political economy of oil in Latin America encompasses two basic groupings, the exporting and importing nations. Although our essays concentrate on the four big countries, any comprehensive survey would follow this scheme: the big exporters—Mexico and Venezuela, followed by Ecuador; and the importers—with Brazil far out in front by volume, followed by Chile, Uruguay, Paraguay, Central America, and the Caribbean. Those nearly or fully self-sufficient—Argentina, Bolivia, Colombia, and Peru—emerged as a third grouping by 1970. In this book, oil-exporting Mexico and Venezuela provide one set of cases, and Argentina and Brazil, which are net importers of petroleum, are historically intertwined in another set of cases.

Mexico and Venezuela arrived at nationalization a generation apart, as Mexico (in 1938) and Venezuela (in 1976) took over large producing oil fields that had been owned, developed, and operated by foreign companies. A glance at the graphs will show quite different export strategies over time. For example, Mexico begins and ends as a major exporter of crude oil. But for forty years crude production closely mirrored consumption; after 1938 the country exported little because it followed an inward-looking strategy of development. Because official Mexican statistics (graph 1) are readily available only for the years since 1938, the decision to resume exporting (based on massive new finds in the late 1970s) seems a startling departure from self-sufficiency. It really is not, given the historical trajectory of the Mexican industry, which began under foreign ownership and continues as the only complete monopoly among the countries under review.

The Venezuelan graph (graph 2), by contrast, shows a continuing export orientation for both crude oil and refined products. Nationalization had little effect on the overall trajectory of the industry;

Graph 1: Mexico: Petroleum Consumption, Imports, Exports, Crude
Production, and Actual Refining: 1910–80 (5-Year Moving Average)

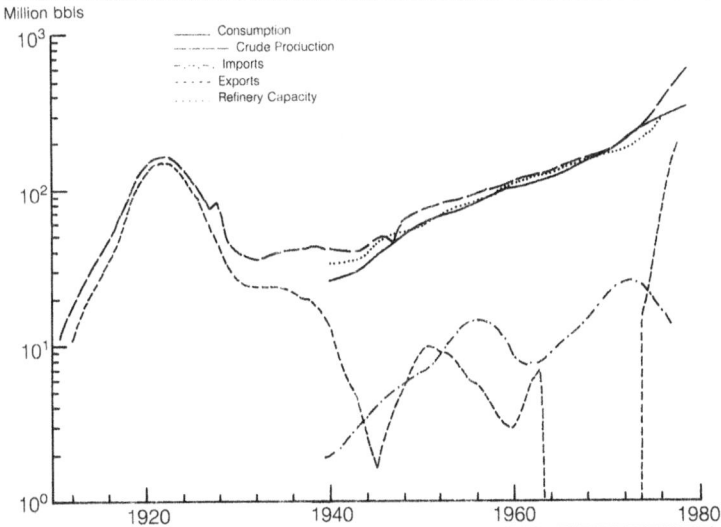

Source: See appendix 1

rather, it completed the policy of gradually tightening national controls that began in the 1940s. The recent trend of rising domestic
consumption in the face of falling production cuts into the nation's
dollar earnings. Today the economies of both countries are heavily
dependent on petroleum-led growth, but the consequences for Mexico are mitigated by the existence of a large, diversified industrial
economy. Venezuela has much less to show for all its export earnings. Ironically, Mexico has now replaced Venezuela as the largest
Latin American oil supplier to the United States. Nationalist policy
has adjusted to growing interdependency between the two North
American neighbors.

Turning to the importers, Argentine oil production soared in the
late 1950s, and since then the country has been on the verge of self-
sufficiency and could become a petroleum exporter (graph 3). The
main reason it has not achieved self-sufficiency is the irregular, stop-
go support for the government company Yacimientos Petrolíferos
Fiscales (YPF), which, as Solberg points out, has been decapitalized
for most of its existence. In fact, Argentina pioneered what are now
known as production-sharing agreements with foreign companies in

Graph 2: Venezuela: Petroleum Consumption, Imports, Exports, Crude
Production, and Actual Refining: 1910–80 (5-Year Moving Average)

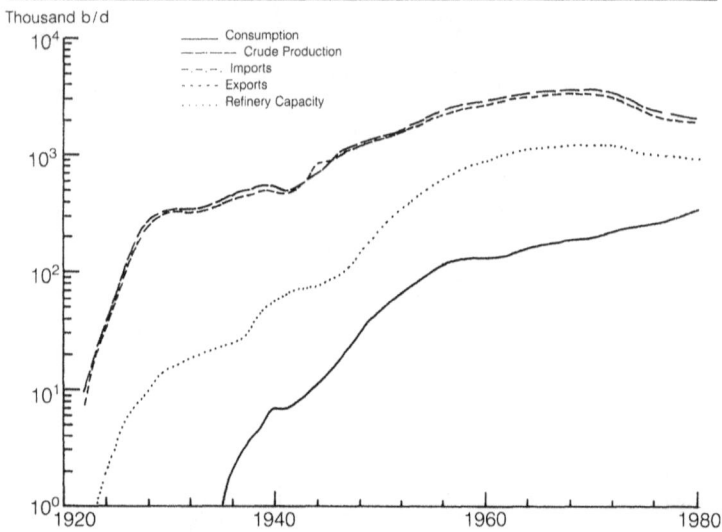

Source: See appendix 2

the late 1950s. Brazil began accepting risk contracts in the late 1970s
to boost investment in exploration. To date, little oil has been found
on this basis, but by 1984 Petrobrás was producing 500,000 barrels
per day of crude oil, more than half the nation's daily requirements
(graph 4). By bringing in expensive offshore fields it even managed to
supplant Argentina as the region's third-largest producer. If YPF is
known as a weak public enterprise, Petrobrás is considered the best-
managed state oil company in Latin America, and it has become a
multinational in its own right. Trading operations account for the
petroleum exports on the Brazil graph.

What links the Argentine and Brazilian cases is the spur that
domestic economic growth, not foreign export markets, provided to
create a petroleum industry after World War I. The nationalist framework that Argentina developed (along with Uruguay) in response to
domestic market signals and that Brazil adopted is covered in the
chapters below.

It is probable that the existence of a strong and growing domestic
market explains why entrepreneurs were attracted to developing

Graph 3: Argentina: Petroleum Consumption, Imports, Exports, Crude
Production, and Refinery Capacity: 1910–80 (%-Year Moving Average)

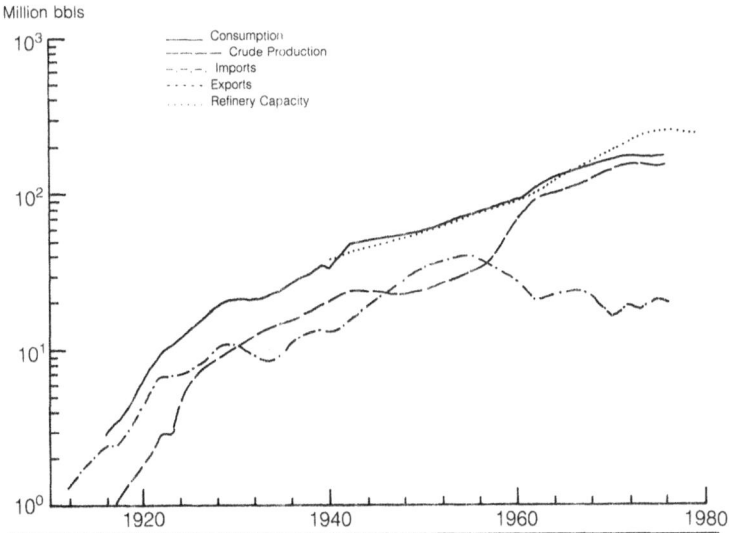

Source: See appendix 3

refineries and sales outlets in the Southern Cone.[3] In Brazil, at least,
domestic capitalists tried their hand at exploration, and not just at
refining and distribution, which were much easier for undercapitalized small companies with a weak technical base to enter. In
contrast to the better-developed business climate in Argentina and
Brazil, Mexico did not attract private national capital to oil before
1938 although, as Durán points out, the government offered it a
role. Beyond selling and speculating on oil leases, this private capital was absent in the overwhelmingly export-oriented industries of
Mexico and Venezuela. (The development of petrochemicals came
later.) Today, the Venezuelan private sector is reported to want a
share of the rapidly growing domestic market in what was until
recently Latin America's leading petroleum exporting economy.

   Brazil is the classic case of what has come to be called the "oil-
importing developing country." Oil was not even discovered until
1939, and into the early 1980s the country was at best a marginal
producer. The fundamental consequences of different resource en-
dowments are illustrated concisely on the graphs, although to be

Graph 4: Brazil: Petroleum Consumption, Imports, Exports, Crude
Production, and Refinery Capacity: 1920–83 (5-Year Moving Average)

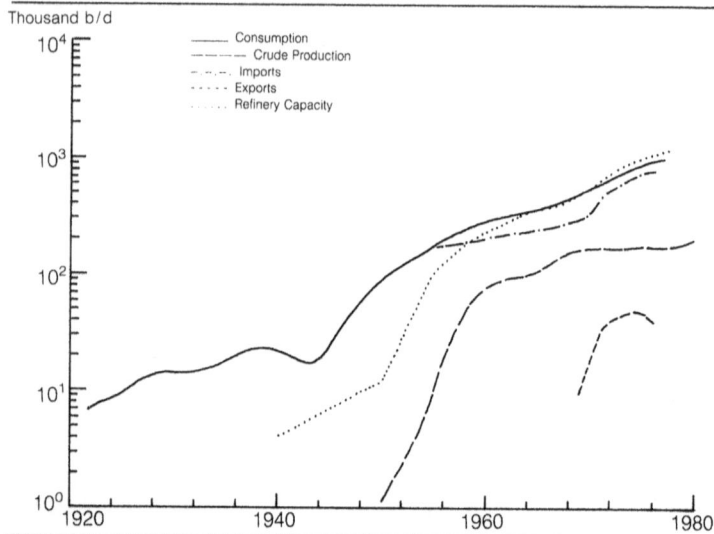

Source: See appendix 4

sure political and economic factors had a great deal of influence on
the actual performances shown here. The Brazilian import bill puts
a heavy strain on that nation's balance of payments. Argentina does
not have this problem, but Mexico, Venezuela, and Brazil all
amassed large public indebtedness in the go-go years of the 1970s to
finance the development of their oil industries. Both of the oil exporting economies overheated, thanks to oil. Venezuela's long-term
failure to invest more of the oil export profits in a diversified economy is well known. It is still unclear how Mexico, where Pemex has
long been an instrument of public policy, will use the company and
its new-found export earnings as a truly strategic resource.

Functions of the Public Enterprise
The state enterprise in Latin America is clothed in a legitimacy and
charged with a breadth of functions that are unimaginable in the
United States, where the state as regulator but not as owner and
producer is the accepted norm.[4] Legitimacy derives from the colonial Hispanic idea of the sovereign's right to regulate, to control,

and to own the economic wealth of the realm. These powers modern Latin American governments have revived for their own ends. Bolstering this tradition of the tutelatory, nurturing state is the body of contemporary codes and statutes that confer upon the nation a broad mantle of legal and above all moral authority to intervene. Laws bestowing broad economic powers within vague time frames vector away from the common law practice of specific precedent and mandated action.[5]

Iberian capitalism featured crown regulatory control of underground resources and of commerce in strategic and popular commodities, such as slaves, mercury, silver, tobacco, and the products of royal properties. In Spanish America, responsibility for carrying out many of these functions was farmed out to private merchant companies, as for example the *asiento*. Portuguese America was less regulated, but in Brazil too capitalists operated within the framework of an interventionist state. Latin America inherited but little of the bourgeois revolution that swept northern Europe, especially England. No clear distinction was made between the political and the economic spheres of society. Modern twentieth-century nationalists have adopted this Iberian heritage for purposes of strategic control and economic growth, creating a climate where the state, not private capital, holds the power. What is new (and still controversial) is their insistence that state industries, not private companies, should operate in these strategic areas.

As they had for centuries, national entrepreneurs were prepared to operate in the shadow of state capitalism, but they asked to be regulated and supported, not crowded out. For their part, the foreign oil companies represented a commitment to free market capitalism of the type that had already been restricted by antitrust legislation in American economic life. As the first multinationals to operate in Latin America, they were highly visible and successful but vulnerable. They were baffled by the willingness and the capacity of apparently weak, pliant governments to circumscribe their economic and legal space. In the public enterprise, both native and foreign private capital confronted a model with sweeping scope and moral force.

Products of a distinct tradition, the state oil companies were also a solution appropriate to nations that were latecomers to industrialization. This is well described by Enrique Saravia, an Argentine specialist in public enterprises:

In underdeveloped countries and, especially, in Latin America, the state transformed itself into the dominant industrial force and the most powerful investor. In addition to providing the operative base for industrialization and development, by constructing infrastructure and providing transport and communication, the state became the only national investor with sufficient economic capacity to implant the industry and develop those basic economic sectors which, for political reasons, could not be left in the hands of the foreign investor.[6]

Their mandate encompasses a broad mix of political and economic goals. In addition to providing basic inputs, the state enterprise is also typically expected to support the overall planning process and to provide employment and other social benefits while turning a profit. These multiple functions are cloaked in national sovereignty, as the International Monetary Fund discovered in 1983, when it asked for cutbacks in Brazil's state sector, including Petrobrás. Thousands of employees marched against this aspect of the fund's austerity plan. Sorting out such broad functions is both a political task and a management task.

Sérgio Abranches, the Brazilian political scientist, has specified what he calls "the inherent ambiguity of the way state enterprises act as they oscillate between their statist face—which shows them fulfilling objectives of a political and macroeconomic nature—and their impresarial face—which shows them favoring discrete interests, goals which could be considered microeconomic."[7] That is, as arms of the state, they are inserted into a set of social and political considerations; as profit-making enterprises they share with private capital bounded goals and the serving of discrete interests. Because of this dual character, the state enterprise draws upon political and economic resources distinct from those that private enterprise can mobilize.

To summarize, "State enterprises are differentiated from private enterprises not only because they are controlled by the state and not always because they are more likely to perform *public* functions, but rather due to their ambiguous or divided identity as economic agents. This ambiguity may be thought of as a structural, distinctive trait of state enterprises, since it derives from their inextricable attachment to two different and sometimes contradictory spheres of activities, each having its own logic of action: the market and the state."[8] Thus in part the functions of state and private enterprise overlap.

Ostensibly, private capital promises a more straight-forward agenda. In a recent manifesto to developing countries, the United States National Petroleum Council said that "competition between independent firms is the best mechanism to bring about efficient resource allocation and foster technological innovation. Private companies are essentially apolitical; their focus is on economic efficiency and technical competence. They are not likely to overlook attractive opportunities in oil-importing developing countries, . . . nor are they likely to squander resources on unsound projects. The OIDCs can benefit by making the fullest possible use of valuable private sector skills and resources."[9] Starting in the 1970s, high oil prices followed by the debt crisis opened up more space to private capital. In the light of history, however, it is probable that as governments come to terms with private capital they will resist and in fact reject the proposition that this industry should be treated only in entrepreneurial terms. However ambiguous, the mix of political and economic tasks the companies perform is rooted in years of practice and experience by the interventionist state. Furthermore, action on more than one policy dimension is probably inevitable in economies of scarcity, where the state must allocate limited resources.

In sum, the multiple roles of the state petroleum enterprise can be grouped into four broad areas:

1. One role of the state industry is to provide an infrastructure that produces inputs for the private industrial sector at attractive prices. That is, the state industry promotes private industry.

2. Another role is to control the commanding heights of a national economy for strategic reasons. To promote national economic self-sufficiency is an older objective, now somewhat out of date. But to integrate plans for military, financial, and development programs into an overall industrial policy planning process is very much on the current agenda for the state companies.

3. Still another role is to promote social development, in some countries by subsidizing energy for consumers and generally by setting an example in the areas of social and labor policy and labor relations.

4. Last but not least is the entrepreneurial role.

Having to perform all four functions has of course led to contradictions. Consider the allocation of profits from operations, a troubling issue for the public enterprise. Profits can be plowed back into operations (Petrobrás is the leading case), used to subsidize other government operations or even to provide general revenues (the case of YPF), or foregone to provide cheap fuel for domestic growth, in effect subsidizing development (until recently the cherished goal of Pemex). The more autonomous and independent the company, the more it will use reinvestment as a way to avoid sharing profits. Until recently, Venezuela's PDVSA (Petroleos de Venezuela, S.A.) was said to be less "politicized" than Pemex, with its post-1938 mandate to provide jobs, patronage, and cheap fuel. That all four nations are deeply in debt today provides an acute perspective on the range of functions their oil companies perform.

For example, after the first oil price rise in 1973 Petrobrás was subjected to an excess profits tax, which went to the central government, although still leaving substantial income to expand operations and pay dividends to some 110,000 shareholders. The United States government imposed a similar tax on the American private oil industry, but Petrobrás has been called upon to do much more, to become a quite flexible policy instrument in times of crisis. To sustain a high level of economic growth, crude oil imports were allowed to rise 8 percent a year until 1979, when the nation faced a payments crisis. Well before the second oil shock, however, Petrobrás was seeking hard currency earnings through Braspetro, its foreign operating subsidiary, and then through Interbrás, a trading company that by January 1983 was providing a wide range of engineering, financing, and commercial services in twenty-four contracts worth $1.7 billion in the Middle East, Africa, Europe, and Latin America.[10] When the debt crisis struck in fall, 1982, the giant company was reported to be stretched thin helping the government solve some of its domestic and foreign problems, using such measures as: income transfers from unfavorable exchange rates; bail-out purchases of alcohol, pig iron, and steel stocks that were then exported at a loss; subsidizing domestic fertilizer prices to make Brazilian agriculture more competitive in the world market; and selling petrochemical raw materials below production costs.[11] Because of the legerdemain of Planning Minister Delfim Netto, Petrobrás was also assigned the role of leasing back from foreign companies off-shore drilling platforms that were manufactured in Brazil, thus contributing to Brazilian "export" targets.

Shouldering burdens not related to the oil industry is unprecedented (to this extent) in Brazil, which now is confronting an industrial depression after nearly three decades of growth rates averaging nearly 7 percent a year. But to the extent that Petrobrás can alleviate extreme disequilibrium in the economy by these means, it is safeguarding the national interest against a payments crisis, thereby fulfilling (however indirectly) a primary reason for its establishment in 1953. Concurrently, the accelerated domestic exploration program is showing results. Whether foreign companies will now assume the main burden of future exploration, something which in the face of budget cuts would enable Petrobrás to concentrate on production, remains to be seen.[12]

PDVSA is caught between weakening export prices, especially for less valuable heavy crudes (30 percent of the total volume of exports), and pressures from within the government to contribute more of its after-tax profits to the country's non-oil development. Since nationalization was completed in 1976, the company has tried to be self-financing to avoid financial (hence political) dependence on the state or foreign investment. Recently, to protect its independence and investment reserves, PDVSA has been exporting beyond its OPEC quota. The president of PDVSA, Gen. Rafael Alfonzo Ravard, in a move to safeguard a guaranteed percentage of oil earnings for reinvestment regardless of the general state of the economy, "advocated that Venezuela put all revenue from oil exports over 1.2 mn b/d into a special fund to repay the country's public debt. The idea of a public debt amortization fund would be to keep the government aware of the need to control current expenditure, at the same time as improving Venezuela's standing in the financial markets, and avoiding the trap that Mexico has fallen into of expanding debt too rapidly on the basis of over optimistic assumptions of future oil revenues."[13] However, the growing debt crisis overtook this strategy: in early 1983 the Central Bank won control of PDVSA's reserves, and Alfonzo Ravard, one of Latin America's best-known technocrats, resigned.

Because of their entrepreneurial role and because they dispose of economic and political resources distinct from those of the state itself, state enterprises cannot necessarily be considered closely controlled instruments of state policy making. The extent to which management is able and inclined to pursue autonomous policies that may even contradict well-established state policies deserves attention. The case studies below show that for many years management at Pemex and YPF have been greatly constrained by state political and economic decisions. Management at Petrobrás and PDVSA, despite their more entrepreneurial thrust, has lost power to the state during the current debt crisis. One of management's biggest problems is having to respond to changing, confused, and contradictory signals from the state, as Saulniers concludes.

The stop-go policies of Argentine governments toward YPF are a case in point. Horacio Boneo claims that left-leaning governments in the early 1970s, seeking low prices for public sector production and high levels of social services of one form or another, tended to conflict sharply with state enterprises. Even though these governments tended to favor expansion of the public sector, their relations with state enterprises were stormy. The conservative military regimes that have ruled for most of the period since 1965 have wanted to contain the expansion of the public sector while making public enterprises run more like private businesses with reasonable profit margins and high degrees of managerial autonomy. As a result, even though hostile to the public sector (except for military-run industries), these governments have tended to maintain cordial relations with their hand-picked public sector managements.[14] The counterpoint theme, as Solberg points out in his chapter below, is that *all* governments since the 1930s have milked YPF for revenues, leaving it an historically undercapitalized state company with weakened management.

The upshot is that the state's relationship to its public enterprise is central to understanding state energy policy in Latin America. According to Jeffrey Seward, "this relationship may be at least as complicated and tortured as the relationship between the state and private companies in the past."[15] Mitigating factors would include the type and degree of consensus built up in support of the state oil companies. In Brazil, for example, there has long been more emphasis on entrepreneurship and a more flexible approach to monopoly than in Mexico, with its union- and private sector–based emphasis on providing jobs and cheap inputs to the economy, goals which historically dovetailed nicely with broad popular support for Pemex. In Venezuela, it is too soon to say whether the entrepreneurial companies formed in the wake of nationalization will retain state backing. Argentina has never had consensus on the goals of YPF.

The Latin American oil companies are among the region's biggest international borrowers, contributing heavily to the current debt crisis. Pemex alone owed more than $20 billion to foreign creditors by 1982. The era of cheap private credit in the Eurodollar markets beginning in 1970, followed by recycled petrodollars after 1973, encouraged borrowing at what amounted to negative real interest rates. Expansionary growth policies were then being cut back drastically by the exporting nations, which had counted on higher world prices, and by the importers. The irony is that oil-rich and oil-poor alike had to accept International Monetary Fund (IMF) austerity programs. The control of natural resources by state oil companies has not led to autonomy. Instead, the "petrolization" of Mexican and Venezuelan economies has made them more vulnerable to cycles of boom and bust. Economic growth rates that averaged 5 to 7 percent since 1940 have almost ceased in Mexico, and in early 1983 the nation faced the near-term prospect of weakening oil prices. Payments for imported petroleum finally restrained Brazilian growth by 1980. If Brazil is now promised some price relief, the fact remains that price swings complicate the conversion to alternative energy sources. Nearly self-sufficient Argentina might be expected to ride out the storms were it not for the divisions in that country over political and economic policy, making YPF probably the least capable of the four big state companies to fill the multiple roles of the public enterprise.

Adverse times throw into sharp relief the strengths and weaknesses of the state oil companies. Once large enterprises in name of the nation have been created, the question of how and for whom the national interest is served is easier to pose than to answer. The state oil companies have been successful in providing externalities for private capital, especially in associated development projects such as petrochemicals; in setting national price structures; and as counterweights to the multinationals. The promise of a generalized rise in living standards for all the inhabitants is still unmet. Also at issue is whether large entities that are comfortable and familiar with closed-door decision making can be made more accountable to the public, which today wants more open societies.

Sheer size propels these companies to assume roles in the international political economy far beyond what the founders intended. In each country we cover, the state oil entity dwarfs any other enterprise. A small company for most of its 45-year existence, Pemex jumped to eleventh place in revenues among all industrial companies outside the United States, while Petrobrás ranks nineteenth.

What is especially interesting in the Latin American industry is the emphasis on integration and cooperation. Because their histories are intertwined, these companies share a pan-Latin vocabulary. This is useful when they are called on to fulfill foreign policy roles.

In 1979, Mexico and Venezuela committed themselves in the Declaration of San José to supply 160,000 b/d to nine Central American and Caribbean countries at what amounts to a 30 percent discount. The trading company Interbrás, a Petrobrás subsidiary and an instrument of Brazil's economically oriented foreign policy, since the late 1970s has supported multilateral ventures in Latin America. Even Argentina, which historically has taken a narrow definition of national resource sovereignty, is now involved in a range of energy programs in the Plata basin.

All four countries have played continental roles in the past with respect to oil, but continentalism is stronger today. A shared regional vocabulary and identity is being channeled by market forces and new policies toward new forms of interdependency. There is a shared perception that greater control of national economies, at least in the energy sector, will not become a reality in Latin America until effective cooperation and integration among the various state companies becomes a reality. This most interesting and challenging thesis harks back to General Enrique Mosconi, who sought self-sufficiency. (The contemporary emphasis on forging an effective industrial policy rather than on older, resource-based strategies nonetheless still draws on this tradition.) To a brief look at the history of Latin American thinking on regional integration we now turn.

Toward Integration

In 1917, Mexico reclaimed sovereign control of its natural resources and abandoned nineteenth-century liberal legislation inspired by the United States mining code, which permitted private capital to own these resources, including oilfields. In reviving the Spanish colonial framework, the Mexicans began a trend that spread throughout the hemisphere. This concept, rather than the 1938 oil nationalization (which was a response to specific local issues), had the greater impact on Latin American economic thought.[16] Other concepts, policies, and experiences were generated in the region and became part of what might be called the idea pool of pan-Latin thinking on petroleum. For example, the policy of assistance by the personnel of one national oil company to another, especially in the early stages of formation, has a long history. (Panama's Hydrocarbons Law was drafted with the assistance of Venezuelan experts and OLADE, the Latin American Energy Organization, to cite a recent example.) Latin American engineers and other professionals perceive a community of interests going back to the early history of the industry.[17]

The exemplary leadership of two public figures, Mosconi of Argentina and Juan Pablo Pérez Alfonzo of Venezuela, set the tone and much of the agenda. Mosconi, the creator of YPF, preached the ideology of national resource control throughout the hemisphere before World War II. Pérez Alfonzo, the founding father of what is now PDVSA and the chief conceptualizer of OPEC, was the apostle of continental cooperation after 1945. Both men were articulate and effective champions for the state company solution.[18]

Less well known is their obsession with efficient management and technical capacity and their insistence that the state corporations had to prove themselves as viable economic enterprises in order to compete with, and replace, the private companies. This message was well received by engineering and technical groups. With import substitution industrialization well under way by the late 1930s, these professional groups were actively promoting Latin American technical solutions and the transfer and sharing of technology. Realizing that the sharing of technology and market information among the multinationals was to their advantage, the engineers devised a counterpart of their own.[19]

The first pan-Latin organization to address these goals was the Instituto Sudamericano del Petróleo (ISAP), which was founded in 1942 and maintained a headquarters in Montevideo until 1951. ISAP was an outgrowth of the Unión Sudamericana de Asociaciones de Ingenieros, a professional group organized in national chapters that held several congresses in the region before World War II. Its statutes and the constitution were written by a small group of petroleum engineers from Argentina, Uruguay, and Bolivia. The driving force behind ISAP was Ing. Carlos R. Vegh Garzón, general manager of the Uruguayan state company, Administración Nacional de Combustibles, Alcohol y Portland (ANCAP). At the time, Vegh Garzón was active as a consulting engineer to refinery projects in Brazil, Peru, Chile, Guatemala, and Colombia. In all these countries, he talked to local engineers about ANCAP as a model of a national industry based on the control of refining.

At first, ISAP was envisioned as a forum to advocate nationalization throughout Latin America, as well as to pool and exchange information and to promote contacts among engineers in the state companies. In a letter laying out the rationale for ISAP in 1940, Vegh Garzón cited the trend to create national companies: it began in Argentina, then spread to Uruguay and Bolivia, and had come most recently to Brazil before continuing on, he thought, to Chile, Colombia, and Peru. ISAP would further the collaboration that

has already been carried out effectively among Argentina, Bolivia and Brazil, the example being given by Argentina which contributed effectively in the development of the nationalistic petroleum policy in Uruguay and Bolivia. More recently Argentina and Uruguay have placed all of their experience at the disposition of Brazilian technicians in order to assist these technicians on the arduous path on which they have started. . . . It is [now] necessary . . . to give firm and definitive form to this collaboration among the South American countries by creating a permanent institute which will collect information on the accomplishments of all the countries in order to place this information at the disposal of those that require it, this being in our judgement the fundamental reason justifying the creation of the "South American Petroleum Institute."[20]

ISAP would also enable engineers who once paid little attention to each other, having "had their eyes fixed, we may say almost exclusively, on the North American petroleum industry," which pioneered this field, to get to know each other and to share technical and marketing information.

Originally conceived as an organ of the *fiscales* (state companies), ISAP was organized in national chapters; the Engineers' Association and the South American Railways Congresses were the first pan-Latin entities of this sort. Each chapter was headed by a government technocrat, although representatives of private national capital (but not the majors) were invited to participate. Uruguay and Argentina were the first to organize national chapters, in 1941; Chile, Peru, Bolivia, and Ecuador joined in 1942, and Brazil in 1946. Vegh Garzón organized the Ecuadorian chapter at the airport on his way back from Washington. However, he was unable to bring in the larger producer countries. Venezuela was not interested. In Bogotá, Standard Oil and the United States embassy urged Colombian officials to set up a Colombia Petroleum Institute, along the lines of the American Petroleum Institute (API), which was open to everyone active in the industry.[21] The upshot is that the fledgling ISAP had to limit its scope to the small producers and the importers.

By mid-1941 the goals of ISAP began to shift toward inter-American cooperation, as wartime shortages produced a crisis. Under Washington's direction, state and multinational oil companies cooperated to allocate scarce fuel through the Petroleum Pool. The prominent role played by the API in setting up the pool and the particularly forceful personality of William R. Boyd Jr. of API, who became the pool's president, were new elements in a much-altered international environment. Vegh Garzón himself worked effectively to bring the South American nations into the pool and to legitimize it. By 1942 ISAP was welcoming local representatives of the international oil companies, and Boyd and his British counterpart were made honorary presidents. Now modeling itself upon the API, ISAP dedicated itself to exchange technical information, hold congresses, publish a bimonthly journal, and explicitly avoid political issues.[22] ISAP thus became an organ to serve everyone in the industry.

By 1950, however, ISAP had clearly failed to become a strong institution; tight travel budgets prevented the national chapters from attending several congresses, many of which were postponed. Also, it seems clear in retrospect that the API model was not readily transferable to South America, which unlike the United States had not yet developed a pool of engineering talent and small companies independent of the majors or the state companies. Furthermore, there was an inherent tension between the fledgling state companies and the multinationals, who were wary of each other. Thus the sharing of technical information through ISAP never went very far.[23]

The inspiration and style of ISAP was quintessentially Uruguayan: a very small country with no power ambitions of its own had been innovating for a generation since José Batlle de Ordoñez began the state industrial projects, out of which came ANCAP in 1931. With no oil of their own and dedicated to *batllismo*, the Uruguayan technocrats and politicians had a continental vision that was cooperative, interactive, and associational. Yet ISAP could not expand its international niche. When asked to associate, both Mexico and Canada said that they already belonged to the API, and

Washington considered regional organizations like ISAP unnecessary. Being close to the oil companies, Venezuela and Colombia
would not join.

There was also an inherent tension between the goals of cooperation and the national objectives of the governments that provided
the bulk of ISAP membership. YPF officials helped ANCAP build
its refinery and were influential in ISAP, but Argentina defined its
national interest in narrow terms. Argentina refused to join the Petroleum Pool, hindered ANCAP's efforts to obtain drilling rights in
Bolivia and Paraguay, and did little to facilitate the riverine transport of Bolivian petroleum, considering this its own area of market
interest. Tensions with the Americans ran high, to the extent that
Washington was prepared to set up an inter-American petroleum
council if Argentina succeeded in dominating ISAP.[24] Later, Juan
Perón may have wanted to use ISAP to promote his own pan-Latin
policies, for it is perhaps not coincidental that ISAP disappeared
shortly after its governing council voted in 1950 to move the headquarters to Buenos Aires.

ARPEL (Asistencia Recíproca Petrolera Estatal Latinoamericana),
the successor organization, was the inspiration of Pérez Alfonzo and
a product of the movement toward integration that swept through
Latin America in the 1960s. When he was Minister of Mines and
Hydrocarbons, Pérez Alfonzo convoked (and financed) the Conference of State Oil Companies of the Americas in Maracay, Venezuela, from June 26 to 29, 1961. Conferees were asked to discuss each
state oil company's administrative and technical problems, areas in
which the private companies were strong. The conference also
focused attention on Latin American petroleum markets, a topic of
concern to Venezuela, which could no longer count on easy access to
the United States after the Eisenhower administration raised duties
on oil imports. Another agenda item concerned ways that the national oil companies could enhance the public welfare and increase
per capita income while promoting industrial diversification and
self-sufficiency, goals Pérez Alfonzo had already established (but not
yet achieved) for Venezuela.

At first ARPEL did not get off the ground. Mexico was to have
convened the second meeting but withdrew; with its nationalist credentials well established, Mexico was evidently wary of Venezuelan
leadership and wondered how a country with a classic export economy dominated by foreign capital could now champion state com-

panies. Apparently Mexico also acted out of caution not to jeopardize its postwar cooperative relations with the American oil companies. YPF hosted the second meeting in 1964, and ARPEL was formally established in Lima in 1965. The statutes and a governing body, including a secretariat in Lima, were approved in Rio de Janeiro in August 1965. And in 1967 the secretariat was moved to Montevideo to be closer to the Asociación Latinoamericana de Libre Comercio (ALALC), the free trade association.

ARPEL is a nongovernmental organization serving the state oil companies, whose employees staff the secretariat and attend the yearly congresses. The aim is to strengthen the state companies,

to study and recommend . . . any measures which would permit mutual collaboration in the defense of common interests with a view toward economic and technical integration. In particular, ARPEL should stimulate the interchange of information and technical assistance among its members; carry out studies to facilitate the development of commercial transactions among members—the expansion of the oil and petroleum equipment manufacturing industries in Latin America, better conservation practices and the coordination of development programs; and promote the holding of congresses, conferences and technical or scientific meetings for the study and debate of matters of common interest.[25]

This broad mandate is reflected in the ARPEL slogan: "Hacia la integración petrolera estatal latinoamericana." Like the old ISAP, it draws on a perceived common heritage of cooperation among technical personnel in the state companies.

ARPEL's emphasis on management training and technical exchange continues the tradition among petroleum engineers begun by Argentina and Uruguay before World War II. ARPEL helped set up the Ecuadorian oil company, Cia. Ecuadoriana de Petróleo and most recently the Paraguayan company Petropar. It also coordinates the sharing of production data, consumption, refining, and market information, all of which used to be dominated by the private companies. The state companies are stronger than they were in the ISAP period. For their part, the majors are said to have established a good record of technical cooperation through ARPEL, where they have observer status. ARPEL also has close links to the API. Canada, which has a long history of significant government management in the national resource field, has joined ARPEL.

Yet to date the organization has not been able to achieve its broader objectives with respect to economic integration. ARPEL moved to Montevideo in 1967 to foster a Latin American common market for petroleum products. But the proliferation of national exchange policies, coupled with volatile international pricing in the 1970s, militated against a common market in petroleum. ARPEL tried to promote a joint engineering company to study, plan, and build petrochemical plants in the 1960s. Efforts to coordinate the manufacture of petroleum equipment in Latin America did not bear much fruit. Recently, ARPEL was studying the feasibility of setting up multinational enterprises among its member companies and voted to establish a Latin American petroleum bank to finance regional oil projects.

The contrast with CIER (Comisión de Integración Eléctrica Regional), the electric power organization established in Montevideo in 1964 and probably the most successful Latin American technical organization to date, is enlightening. Like ARPEL, CIER promotes technical cooperation among public companies. It has successfully promoted standardization and the interconnecting of a power grid among several South American nations. CIER has also become a sector-wide agency for South America: while more than 80 percent of the firms are public enterprises, a surprising number of electric cooperatives and even some private firms belong to the organization. Given the technology and the product, electric power is much easier to coordinate than petroleum.[26]

If the sharing of technical, marketing, and management information is well established, divisions over political and economic policy have lessened the effectiveness of ARPEL. For several years Mexico stayed out of ARPEL, apparently fearful that it would become a Latin American OPEC with links to the cartel. This was never intended. Pemex did join later on, and Venezuela after the initial impetus did not dominate ARPEL. Instead, Uruguay became the principal organizer and promoter. Uruguayan leadership suffered, however, with the decline of ALALC, the proliferation of different exchange policies, the rise of bilateral economic relations in the mid-1970s, and above all the strong trend to extreme economic liberalism sweeping through Chile, Argentina, and Uruguay. Until the recent failure of the so-called Chicago Boys (free marketers from the University of Chicago school of economics), it was difficult and even dangerous for engineers to say anything good about the state company model.

The state companies are available instruments to promote regional integration, but forward movement has been slow. Only in 1976, when Venezuela completed its nationalization program, did the Latin American state companies control a clear majority of the region's oil production. Until then, none of the large state companies was prepared to finance operations outside its own borders. Venezuela excepted, their ideology was to serve the internal market. This began to change in the face of rapidly rising energy costs, especially after the second oil shock in 1979.

Braspetro, for example, signed a production service contract with Colombia in 1976, but almost all of the Brazilian overseas efforts had been concentrated on Africa and the Middle East. The growing Latin American markets for its manufactured goods and the search for less politically vulnerable, more dependable and available oil supplies explain why Brazil developed a strong Americanist foreign policy. For Brazil, the binational agreement with Paraguay to build the Itaipú hydroelectric project became something of a model for joint operations. Argentina and Brazil ended their bitter dispute over competing hydroelectric projects in 1979. When Brazilian oil imports were disrupted by the Iran-Iraq war, both Mexico and Venezuela offered to help although neither had been suppliers to Brazil for many years. Thus when Venezuela proposed a tri-state operating company with Mexico and Brazil in 1981, the Brazilians agreed. Planning began for Petrolatín, a new multinational company with capital and equipment subscribed by the three state companies. The first president will probably be a Brazilian, and if Petrolatín ever gets off the ground it will be the first Latin American transnational. In the words of Venezuela's Humberto Calderón Berti, the minister of energy and mines who championed the concept, Petrolatín would be an increasingly powerful "instrument to encourage their economies and to strengthen their negotiating capacity with the rest of the world."[27]

(At current writing Petrolatín is postponed and may well be dead because of the debt crisis. Since the money ran out in mid-1982, moreover, the IMF has cautioned against expansionary activities in the state sector. Instead, the three nations are signing clearing agreements that include petroleum and oil field equipment. Following Mexico's lead, the four large debtor nations are cooperating and, as of April 1984, may coordinate their debt negotiations in order to gain leverage in the international financial system.)

Venezuela took the initiative with Petrolatín after first approaching the Carter administration about United States government involvement in a multinational project. The objective was to accelerate production and alleviate the combination of acute oil shortages and high prices that were strangling regional economies after the second oil shock. Secretary of Energy James Edwards was invited to Caracas, and discussions with the majors were held as well. Carter and his beleaguered administration were cool toward this initiative; President Reagan with his cost-cutting, free market principles was against it.

Mexico at the height of debt-led growth was also eager to widen the scope of Pemex' activities in the region. "To sustain a certain rate of growth without pressing on the desired volume of domestic production," Jaime Corredor observed, "it may be expected that the Mexican petroleum industry, together with various collatoral sectors, will tend to go abroad to cooperate with other countries in the exploration and development of hydrocarbons, or to participate on contract in the construction of petrochemical plants and refineries. This may open the door for a new type of cooperation between Mexico and some foreign companies—even U.S. companies—to operate in third countries (through joint ventures), although it may also signify a new field of competition with the United States."[28] (This scenario vanished with the debt crisis. Temporarily, at least, Mexico stepped up sales to the United States petroleum reserve as Reagan cooperated closely with the new Miguel de la Madrid administration on the debt.)

At the same time, Mexico and Venezuela were developing a joint energy policy for the Caribbean basin. Mexico under President José López Portillo was searching for influence and markets in Central America and the Caribbean after the overreaching policies of his predecessor, Luís Echevarría, who championed grandiose integration schemes like the Latin American Economic System (SELA) while seeking and failing to achieve world leadership. Venezuela perceived an opportunity to exert leadership as a rising middle-range power in the Caribbean basin. Both major exporting countries were at pains to demonstrate that the oil price rise had not elevated them to the company of rich exploiting nations. In the Declaration of San José, Costa Rica, in 1979 they pledged themselves to supply 160,000 b/d to nine Central American and Caribbean nations (excluding Cuba) at concessionary prices. (Even after the debt crisis struck in 1982 they maintained this assistance, which went beyond anything that Washington was yet prepared to offer the Caribbean basin. By 1984, however, both had cut off oil shipments to Nicaragua.) The Venezuelans also wanted badly to show that something could come out of OLADE, the Organización Latinoamericana de Energía, which since its inception in Caracas at a meeting of energy ministers in 1972 had accomplished little. OLADE called the emergency meeting into session at San José in July 1979 to deal with the inability of most Central American and Caribbean nations to cope with the energy crisis.

OLADE, the third Latin American organization to deal with petroleum, brought together the ministers of mines and energy from several countries to discuss overall energy policy.[29] Like SELA, however, OLADE was conceived with larger goals than could realistically be achieved in the 1970s. Its most ambitious goals included multinational energy companies, pooled financial instruments, and a Latin American energy market. Only Venezuela could have financed the lion's share of these projects, and that nation's development planning was overwhelmingly focused on national needs, despite the pre-1979 rhetoric of regional economic planning and transnational enterprises. Furthermore, a minister might or might not have effective control over his national oil company, depending on the degree of autonomy. Until the late 1970s, for example, Petrobrás was accustomed to making what amounted to foreign policy decisions with little direction from the minister of mines. OLADE's ineffectiveness can also be explained by the fact that Ecuador, which hosted the secretariat, was in turmoil over its own energy policy. This imposed a political and ideological cost on OLADE.

Venezuela took the lead with Mexico and Brazil to establish in Caracas in October 1981 the Latin American Energy Cooperation Program (PLACE). According to OLADE, "PLACE seeks to integrate regional energy production with socioeconomic development and cooperation in order to broaden and diversify Latin American energy production."[30] At the Caracas meeting, Brazil, Mexico, and Venezuela declared their intention to join forces for oil exploration, and planning for Petrolatín began.

Whether grandiose or modest, all these projects for integration must be steered around the rock of national sovereignty in a region known for fragmentation and divisions. OLADE signatories believed that "natural resources can be an element of regional integration [and that], furthermore, as an official organization of Latin American countries, OLADE could facilitate this integration without usurping each government's sovereignty over its natural resources." This credo was more than Argentina, which did not ratify the Lima Convention setting up OLADE in 1973, was prepared to accept. Argentina, which had pioneered the doctrine of resource independence under Mosconi, did join OLADE eight years later "with the understanding that such ratification does not in any way alter the traditional position of Argentina in the field of shared natural resources. . . ."[31] However, this position was overtaken by the logic of integration. In addition to importing Bolivian gas, that is, from its traditional economic heartland, Argentina was preparing to supply Brazil with natural gas until the Falklands War in 1982 upset a construction timetable that was already well advanced. Argentina spurned a Brazilian proposal to build a joint satellite communications system; it was unwilling to subsidize wheat shipments to Brazil in order to compete with traditional suppliers; it had no desire to become a neocolonial supplier of raw materials to its more powerful neighbor that was now an industrial power. In turn, Brazil spurned an Argentine proposal to integrate the electronic industries of both countries. This was another consequence of the Falklands war, which revealed how fragmented the Latin American nations still are. Yet for both Brazil and Argentina the Plata basin is a promising growth pole, and petroleum will play an important role in a growing regional market.

OLADE also accompanied the return of private capital into exploration and production on new terms and conditions during the 1970s. With many countries beginning to sign the new risk contracts, OLADE established an advisory board to pool information and to provide support to member states. OLADE was said to be counting heavily on the cooperation of ARPEL, which had its own advisory service on the risk contracts. In 1982, the two organizations agreed that ARPEL would assume responsibility for the hydrocarbons sector.

The age of easy oil discoveries was probably over; the era of capital shortages was arriving. Even countries with strong technical resources, like Argentina and Brazil, no longer had the economic capacity to go it alone. The "concern for property [rights], sovereignty, public service, economic and social development, national security, and other important factors has led many Latin

American countries to independently explore and exploit their own hydrocarbon resources, and to exercise more effective control over their operating companies," one observer commented in 1981. "Except for Venezuela and Mexico, [however], Latin American countries have signed or are going to sign risk contracts."[32] Two years later Venezuela was under pressure to reexamine its exclusionary legislation, and in all of the countries surveyed in this book the domestic private sector had become more active in petroleum.

Afterword

To historians, the present is prologue to the past. I hope this brief excursion into contemporary issues has established certain propositions about the multiple, if ambiguous, roles of the state companies, while tracing the line of continentalism among the technocrats and managers. These dimensions should not be lost on those whose interests lie primarily with the economics of the oil business. The struggle to achieve policy coherence as well as effective operating companies includes the conflict between private and state capital; but there has always been more to the story than this, perhaps in our century the oldest and most celebrated of confrontations between powerful private entities and the state. To the case histories of this business in the formative years—in essence, to its *historicity*—we now turn.

Notes

Abbreviations

ARPEL	Asistencia Recíproca Petrolera Estatal Latinoamericana
DS	State Department Archives, in United States National Archives
EIU	Economic Intelligence Unit
ISAP	Instituto Sudamericano del Petrólco
OLADE	Organización Latinoamericana de Energía

1. These aspects are well covered in George Philip's excellent comprehensive survey, *Oil and Politics in Latin America: Nationalist Movements and State Companies* (Cambridge and New York, 1982). The author wishes especially to thank Isabel Lussich, who greatly aided his research in Uruguay, and Ing. Carlos R. Vegh Garzón, who provided many insights.

2. This line of thought was suggested by Paul H. Frankel, "The Rationale of National Oil Companies," in *State Petroleum Enterprises in Developing*

Countries (New York: United Nations, 1980), p. 5. For a good case study of how costs and benefits were evaluated—including such aspects as fairness, predictability and control—in contractual discussions with a multinational energy company see Harvey F. Kline, "The Colombian Debates about Coal, Exxon, and Themselves," *Inter-American Economic Affairs:* 36, no. 4 (Spring 1983).

3. Literature on the Latin American private sector in the twentieth century is weak. For leads on the oil industry consult Philip and the chapters below.

4. "One of the best-kept secrets of the United States system is the close to 7,000 public enterprises in this country," Alfred Saulniers observed (personal communication). For details, see Annmarie Hauck Walsh, *The Public's Business: The Politics and Practices of Government Corporations.* (Cambridge, Mass., 1980 ed.).

5. Jonathan Brown provided many insights. For examples of the clash of legal systems see Esperanza Durán's chapter.

6. Enrique J. Saravia, "A teoria geral da administração e a reforma administrativa," in *Textos da disciplina administração e política empresarial das empresas estatais* (Rio de Janeiro: Fundação Getúlio Vargas, CIPAD, [1976?]), p. 76.

7. Sérgio Henrique Abranches, "A questão da empresa estatal—economia, política e interesse público," in *Textos da disciplina,* pp. 197–200.

8. Sérgio Henrique Abranches, "Patterns of Public Entrepreneurship in Brazil" (Paper presented at the 10th World Congress of the International Sociological Association, Mexico City, August 16–21, 1982), p. 5.

9. National Petroleum Council, *Third World Petroleum Development: A Statement of Principles* (Washington, D.C., 1982), p. 23.

10. ARPEL, *Notícias* (Jan., 1983), p. 3. Braspetro was created in 1972; the trading company Interbrás was spun off in 1976 to take market advantage of Brazil's large purchases of crude oil.

11. "Petrobrás Bails out Weak Sectors . . . ," *Latin American Regional Reports, Brazil* (Jan. 7, 1983), pp. 6–7. This report mentions subsidies for other hard-pressed sectors, including the National Coal Program, Transport Fund, air force, merchant navy, and nuclear program.

12. EIU, "Petrobrás Leaves Exploration to Foreign Contractors," *Quarterly Energy Review, Latin America and the Caribbean* 1 (1983): 22–23.

13. Frank E. Niering Jr., "Venezuela, Doubts over Long-term Objectives," in *Petroleum Economist* (March 1982), pp. 85–87. The quote is from Economist Intelligence Unit, *Quarterly Energy Review, Latin America and the Caribbean* (4th Quarter, 1982), pp. 20–21. The EIU's *Annual Supplement, 1982* comments further: "The extent of the oil industry's support for the national economy can be gauged by the fact that 95 per cent of export sales are derived from oil and 75 per cent of government revenue is provided by taxes

on the oil company" (p. 84). On the company's resistance to raising production to satisfy short-term financial crises consult Philip, p. 124.

14. Horacio Boneo, *Political Regimes and Public Enterprises*, Technical Paper Series 31 (Office for Public Sector Studies, Institute of Latin American Studies, University of Texas at Austin, 1981).

15. Political scientist Jeffrey Seward made this point to me, adding that "there is *both* an economic-political tension in the behavior of state enterprises *and* a public-private tension. To some extent the state defines the economic and political agenda of the enterprises, but to some extent the enterprises define their own political and economic agendas and, under some circumstances, may even be able to impose their semiprivate agendas on state policy rather than the other way around." Among the four state oil companies under review, Petrobrás (historically the most "independent") would provide the best test for his hypothesis.

16. Gen. Júlio Caetano Horta Barbosa's library in Rio de Janeiro contains a bound copy of the Mexican Mining Code and the 1917 Constitution with the famous Article 27, reserving ownership of the subsoil wealth to the nation. This volume is annotated in the hand of Colonel Juarez Távora, who as a young reformist officer set up Brazil's Mining Code of 1934.

17. Interview, Ing. Carlos Vanrell Pastor, an ISAP activist and later in the ARPEL Secretariat, in Montevideo at ANCAP headquarters, Oct. 13, 1982. The assistance for Panama's new legislation to regulate the exploration and production of hydrocarbons is reported in ARPEL, *Notícias* (July 1982), p. 10.

18. Mosconi's dates are 1877–1940; Pérez Alfonzo died in 1979. For an excellent brief summary of Mosconi's career and legacy consult Carl E. Solberg's *Oil and Nationalism in Argentina: A History* (Stanford, 1979), and for Pérez Alfonzo see Franklin Tugwell, *The Politics of Oil in Venezuela* (Stanford, 1975). New information on these key figures is also provided by Solberg and Lieuwen in the chapters below.

19. After World War II, the sharing of technology among State companies spread to other regions. See M. Nezam-Mafi, "Cooperation among State Petroleum Companies," in *State Petroleum Enterprises in Developing Countries* (New York: United Nations, 1980), pp. 61–76.

20. Letter, Vegh Garzón to J. L. Buzzetti, president of the Unión Sudamericana de Asociaciones de Ingenieros [1940], and published in the Unión's *Revista de Ingeniería* (June 1940). It was translated and sent to Washington from the United States Embassy in Montevideo, Oct. 28, 1940 (DS, Report 642, 833.6363/143). The National Library in Montevideo has a complete collection of ISAP's *Boletín*, published from 1943 until 1951.

21. While Vegh Garzón was sounding out professional groups in Argentina and elsewhere, Standard Oil alerted its subsidiaries (letter, F. C. Schultz of Standard Oil of Argentina to C. H. Lieb in New York, Buenos Aires, Aug.

8, 1940). The new Colombian Institute to share information, lawyers, and technical information "would be a constructive influence in eliminating any ideas of nationalization of the oil properties in Colombia" (memorandum, including the Schultz letter, from ambassador in Colombia to the secretary of state, Bogotá, Oct. 16, 1940. DS 821.6363/1368).

22. *Boletín del I.S.A.P.* 1, no. 1 (April 1943) to the last issue, 3, no. 4 (August 1951). Based also on a long report by Ambassador Baptista Luzardo in Montevideo to the Brazilian Foreign Minister (Ofício 3221, Sept. 23, 1941, in Montevideo Ofícios, Sept.–Dec., 1941, Arquivo Histórico do Itamaraty).

23. Interviews with Vegh Garzón, Oct. 6 and 12, 1982, and with Carlos Vanrell, Oct. 13, 1982, in Montevideo.

24. Memorandum, Laurence Duggan to A. A. Berle, July 5, 1943 (DS 563/63D1/1 GS AW).

25. The quote is from Aníbal R. Martínez, *Our Gift, Our Oil* (Vienna, 1966), pp. 183–85, who provides a brief sketch of ARPEL's formation. See also ARPEL, "17 años de ARPEL," in *Notícias* (Sept. 1982), pp. 2–3. Vanrell provided many insights in the interview cited above. Consult his article: "Latin American State Petroleum Enterprises and their Association in ARPEL," in *State Petroleum Enterprises in Developing Countries* (New York: United Nations, 1980).

26. For a brief description of CIER, consult Eduardo White, *La acción internacional de las empresas públicas en América Latina* (Mexico City: Fundación Getúlio Vargas, Escuela Interamericana de Administración Pública, 1976), p. 37. White also has a brief account of ARPEL on p. 44, with data on Latin American technical agreements to date. The Asociación Latinoamericana de Ferrocarriles (ALAF) was set up in 1964, p. 33. Notice of the Banco Latinoamericano del Petróleo project, which was approved at the thirteenth meeting of ARPEL in Quito, July 17–21, is in INTAL, *Boletín de la Integración* (Sept. 1972), p. 545.

27. José Andrés Oteyza, secretary of industrial development (Mexico), César Cals de Oliveira Filho, minister of mines and energy (Brazil), and Humberto Calderón Berti, minister of energy and mines (Venezuela), "Joint Declaration" [Oct. 16, 1981] in Ministry of Energy and Mines, *Weekly Letter* (Caracas) 24, no. 42 (Oct. 23, 1981). Calderón Berti's rationale for Petrolatín: "Brazil, Mexico and Venezuela are in a position to build a company which could equal other transnationals and which could compete advantageously with them in the field of exploration and development of prospective areas" (Ibid. 24, no. 51 [Dec. 23, 1982]). Both of these documents are courtesy of Lagoven, S.A. Interview with Enrique Saravia, Fundação Getúlio Vargas, Rio de Janeiro, Nov. 8, 1982, and his letter to me, Rio de Janeiro, Oct. 26, 1983. See also EIU, *Quarterly Energy Review, Latin America and the Caribbean: Annual Supplement, 1982* (London, 1982), pp. 1–3.

28. Jaime Corredor, "The Economic Significance of Mexican Petroleum from the Perspective of Mexico–United States Relations," in *U.S.–Mexico Relations: Economic and Social Aspects,* eds. Clark W. Reynolds and Carlos Tello (Stanford, 1983), p. 163.

29. Luiz Augusto Marciano da Fonseca, "The Latin American Energy Organization (OLADE)," *Energy* 6, no. 8 (1981), provides a good description of OLADE.

30. OLADE, *Actualidad energética latinoamericana* 52 (Oct. 1981), p. 1.

31. See Fonseca, "Latin American Energy Organization," p. 861. Argentina's policy toward OLADE is summarized in OLADE's *Actualidad energética latinoamericana* 53 (Nov.–Dec. 1981), p. 3.

32. Fonseca, "Latin American Energy Organization," pp. 863–64; see also ARPEL, *Noticias* (March 1982), p. 2. The two organizations signed an agreement giving ARPEL responsibility for the oil and gas.

1. Mexican oil workers using pipe tongs to assemble pipeline near Tampico, about 1920. Courtesy M. Philo Maier.

Jonathan C. Brown

Jersey Standard and the Politics of Latin American Oil Production, 1911–30

Today, the Exxon Corporation, one of the largest transnational corporations in the world, participates in the development of energy resources in Latin America as one of the many private companies providing capital, marketing, and technology to the state oil industries. Its involvement in the region has a long and sometimes turbulent history.

In the late nineteenth century, the Standard Oil group was the first oil firm to export North American petroleum products to the developing Latin American countries. It built several refineries throughout the region that processed imported United States crude oil, and the oversight of these and much of Latin American marketing became the province of Standard Oil Company (New Jersey), a holding company within the Standard group and principal predecessor of Exxon. Latin America eventually contributed nearly two-fifths of this company's sales profits. Before long, a number of pioneer oilmen began to erode Standard's sales domination by developing Latin America's own oil production in Peru, Mexico, and, combined with a fledgling government producer, Argentina. The United States Supreme Court's dissolution of the Standard Oil group in 1911 stripped the New Jersey firm of its domestic crude oil supplies and forced a decision: either develop production abroad or risk loss of sales in Latin America altogether. Thus began a concerted effort by the New Jersey company to acquire oil fields close to its markets.

Between 1900 and 1930, Latin America was the scene of the world's most intensive oil development outside the United States. Jersey Standard entered into direct production there rather late, having acquired no producing assets until 1913. By that time, Latin

American politicians were beginning to impose restrictions on the foreign companies that sought territorial control of oil fields by changing the terms under which the existing oil companies were operating and by imposing stricter terms for new entrants.[1] Under the circumstances, Jersey attempted exploration activities in Argentina, Bolivia, and Venezuela but succeeded only in the last. But in Peru, Mexico, and Colombia it entered as a purchaser of established producing properties and of existing concessions, in effect buying the generous contracts made in the previous decade. By 1930, Jersey Standard had established itself as the largest producer in the Southern Hemisphere. Its Latin American operations had become so important that they contributed approximately 40 percent of the company's worldwide production of crude petroleum.[2]

Business success, however, did not translate automatically into political triumph. In particular, its production properties—oil fields, pipelines, and the contracts to operate them—drew considerable political opposition. The marketing activities of the foreign oil companies represented a failure to control national resources and attracted little public approbation. Mere public discussion exposed the company to growing criticism as a monopoly and as inimical to national interests. In Mexico and Argentina, two countries that already had a long history of foreign investment, nationalist opposition to production enterprise was especially damaging. Yet even Jersey's dominant position in the oil fields of the less nationalistic countries of Peru and Colombia drew efforts to modify and to cancel the company's production rights. Nationalist attacks, even when unsuccessful, at least had the effect of raising taxes and preventing Jersey's acquisition of additional production in these countries. Jersey Standard was politically safe only in Venezuela, a country that was to become the core of the company's operations in Latin America due not only to its oil potential but also to Venezuela's late developing economic nationalism.

The widely held conception that large multinational companies enjoyed unusual political advantages in Latin America must be treated with scepticism. Indeed, the reverse may be more accurate: its prominence in the economy permitted the multinational corporation very little political manueverability. The Standard Oil Company (New Jersey) is an appropriate example. A worldwide vertically integrated company that had gained a large measure of economic success in Latin American oil development, this firm in the long run was unable to protect, let alone perpetuate, its own private enterprise model. Even before the Great Depression, Jersey was susceptible to the political and administrative requirements of host governments. United States diplomacy offered short-term protection at best. The oil company, therefore, was forced to settle for ad hoc, stop-gap solutions in order to delay the inevitable erosion of its production assets. After 1930, the nationalists won and in country after country succeeded in removing all Latin American oil fields from Jersey's control.

This chapter will analyze the successful entry of the Standard Oil Company (New Jersey) into Latin American petroleum production before 1930 and the early political ramifications of its producing enterprises in Peru, Mexico, Colombia, Argentina, Bolivia, and Venezuela. Brazil and other countries are excluded from this chapter. Although Jersey Standard had an extensive and essentially uncontroversial marketing network in Brazil before the depression, it appeared uninterested in acquiring oil fields there. Some Standard leasing in the north convinced Brazilian nationalists that the company had designs on the entire Amazon basin, but fear and fact were far apart.

Despite the nationalization of Jersey Standard's assets in Bolivia, Mexico, Cuba, Peru, and Venezuela, the present activities of the Exxon Corporation in Latin America are by no means insignificant. Its interests in the Southern Hemisphere are overseen by two affiliates: the Exxon International Company with headquarters in New York is charged with the acquisition and transport of crude oil supplies, and Esso Inter-America, Inc., of Coral Gables supervises the exploration, refining, production, and sales in the region.[4] What has changed over the last five decades is the extraordinary control that governments have extended over all aspects of the oil business, prompting Esso Inter-America to employ political affairs experts as well as engineers and systems managers.

Exxon's current sales in Latin America account for a mere 10 percent of its total worldwide sales of petroleum products, but the marketing complex nonetheless is impressive in its expanse. Esso's two Argentine refineries, the oldest one at Campana with a capacity of 90,000 barrels per day (b/d) and a smaller plant at Puerto Galván near Bahía Blanca, supply a full range of products for more than 800 service stations in that nation. Some crude petroleum is imported but most is acquired from Yacimientos Petrolíferos Fiscales

(YPF), the state oil company. In Chile, the company's distribution network consists of 12 marketing terminals, 167 service stations, and 9 aircraft refueling facilities. Esso's refineries in Nicaragua and El Salvador provide products for 316 service stations in Central America, and its Panamanian Division supplies 17 percent of the bunker fuel for ships passing through the Panama Canal. In Nicaragua, where relations between the company and the Sandinista regime are described as pragmatic, Esso refines crude oil acquired by the government directly from Mexico and Venezuela on concessionary terms and distributes the products at prices established by Nicaraguan officials. Esso supplies nearly 20 percent of the Caribbean area's demand for petroleum products with a chain of more than 800 service stations. Brazil is the company's largest market in the region, and 53 distribution centers and some 3,000 service stations carry Esso products to Brazilian consumers. In addition, Exxon operates petrochemical plants in Argentina, Chile, Brazil, Aruba, and Nicaragua.

Where Jersey Standard properties were nationalized (Bolivia, Mexico, Cuba, Peru, and Venezuela), Exxon today retains few assets. In fact, the company's dearth of direct production is the most distinctive feature of its activities in Latin America (see table 1). It had

Table 1: Operations of Exxon Corporation
in Latin America[a]
(expressed in thousands of barrels daily)

	1977	1978	1979	1980	1981	1982	1983
Petroleum product sales	488	510	505	472	459	453	436
Refinery crude oil runs	368	406	370	402	401	375	341
Net production of crude oil	16	14	13	11	11	10	14
Crude oil acquisitions[b]	825	645	647	432	361	n.a.	n.a.

[a]Latin American data are listed as "Western Hemisphere other than the United States and Canada."

[b]Crude oil acquisitions represent purchase of Venezuelan crude only.

Sources: Exxon Corporation, *1981 Annual Report* (New York), p. 51; Exxon, *1982 Annual Report*, p. 44; Exxon, *1983 Annual Report*, p. 48; Exxon, *Financial and Statistical Supplement to the "Annual Report,"* p. 46.

been lifting Mayan crude oil under contract with Pemex, Mexico's national petroleum company, until the current oil glut forced suspension in July 1981. Venezuela took over its petroleum industry on January 1, 1976, and Exxon's affiliate, the Creole Petroleum Corporation, became the government company, Lagoven. Since then, Exxon employees have provided technological expertise to Lagoven under technical assistance contracts, and large amounts of Venezuelan crude were purchased for the 300,000 b/d refinery on Aruba, closed recently.

More than other countries in Latin America, Colombia has retained historically close and mutually beneficial relations with the giant multinational. In the early 1950s, Jersey Standard turned over the De Mares oil fields, the refinery, and the pipeline to the national oil company in the first peaceful and contractual transfer of private oil assets to a Latin American government. Esso Colombiana still maintains an active marketing apparatus throughout the country and until recently had been drilling exploratory wells. Moreover, the company has formed a partnership with Colombia's state energy corporation in a $3 billion coal mining project in a Guajira peninsula that is expected by 1986 to begin exporting coal to Europe. With contracts already signed for the delivery of coal to Ireland, Spain, and Denmark, the joint venture is engaged in constructing a coal port at Bahía Portete, a 100-mile road, and a railway across the peninsula. Estimated to continue as a joint venture for 30 years, the project will employ 7,500 people during construction and 5,700 when the mining operation reaches full capacity in the early 1990s. Creation of economic activity in a sparsely inhabited area is reminiscent of Jersey Standard's earlier development of the oil industry in the Magdalena valley as well as in northern Peru and in eastern Venezuela.

Facing domestic economic crises brought on by rising petroleum prices after 1973, many oil-importing nations of the hemisphere are reversing their efforts, begun earnestly in the 1920s, to restrict the activities of the foreign companies. State monopoly of the industry is no longer held to be an unambiguous good. At the same time, the Latin Americans clearly intend to multiply the number of private firms and to reduce the power of the largest corporations. In Peru, where Standard Oil Company (New Jersey) once had been the largest producer, the state recently has granted contracts to Occidental, Belco, and Shell. Chile has made arrangements with Atlantic Richfield and Amerinda Hess for offshore drilling. Texaco and

Hispanoil operate in Guatemala, while Occidental and other independents have received Bolivian exploration rights.

Despite recently falling prices of crude, the oil-importing nations of Latin America have not de-emphasized the exploratory programs that several nations had begun with the foreign oil companies, including Exxon, after the 1973 price rise. These countries are heavily in debt and view domestic production of fuel resources as an essential solution to their balance of payments problems. Brazil has granted Exxon several offshore exploratory blocks on the Atlantic shelf in the Santos basin off Bahía and in the Amazon delta. Moreover, Brazil's national oil company, Petrobrás, has agreed preliminarily to assign Esso two exploratory tracts onshore in the state of Maranhão—a departure from Brazilian policy to date that has reserved onshore exploration for Petrobrás alone. Some of Argentina's offshore expectations were dashed early in 1982, when Esso pulled out of its contract area after a thirteen-well exploration effort resulted in only two discoveries of noncommercial oil deposits in the Malvinas basin. Ironically, the company had been using the YPF-owned drilling rig *General Mosconi*, named for the state company's first leader who so tirelessly had battled to reduce Standard Oil's activities in Argentina. While it has gained no such exploratory rights in Chile, Exxon in 1978 purchased the Disputada copper mines for $104 million from the Chilean government. Even though production has increased from 23,800 metric tons of copper in 1978 to 39,200 tons in 1981, falling world copper prices and economic recession have resulted in significant losses.

The contracts that Esso now has in Latin America, differing greatly from the liberal concessions under which its predecessor first entered the region, reserve paramount control to national governments and to state energy companies. Exploration in Brazil and Argentina is undertaken with the risk contract, in which the company has four to six years in which to explore the tract. Should a commercial discovery be made, the private company develops production for sale to the state oil companies at prices indexed to the international oil market. In such cases, YPF and Petrobrás would determine final disposition of the crude oil. Esso operates in Colombia under an association contract that gives Esso exploratory rights for two to four years, and if commercial resources are discovered, a joint venture is established with a state company that contributes half the investment and takes half the profits. Actual Colombian returns more nearly approximate 80 percent, say Esso officers, because the government first collects a royalty of 15 percent on all production.

Such command of their own energy resources has resulted from years of work by several generations of Latin American economic nationalists—whether for good or for ill. It is well recognized today by all parties, for example, that energy resources in the hemisphere are public domain and not the private property of foreign companies. Such government control, a phenomenon steeped in history, can be understood in its novelty if one considers the early operations of the foreign oil companies and, in particular, those of Jersey Standard.

Latin American Markets and Jersey Standard's Style of Operation

While it was a latecomer to production, Standard Oil Company (New Jersey) nonetheless can be considered a pioneer marketer at home and abroad. Indeed, growth of the original Standard Oil group late in the nineteenth century came in refining and sales, particularly in the northeastern United States. Even before the retirement of John D. Rockefeller in 1895, the company had established a refinery at Bayonne, New Jersey, and led all American companies in exporting petroleum products. The United States, then the world's leading supplier of petroleum products, exported $50 million of petroleum products in 1889, of which Standard Oil accounted for 90 percent.[5]

Within the larger Standard Oil group, the New Jersey holding firm controlled the Bayonne refining and transport complex as well as a growing share of the group's foreign sales. Considerable refining abroad was also undertaken in order to serve its Latin American markets more efficiently. Jersey established refining plants in Mexico City and Vera Cruz after purchasing a majority interest in the Waters-Pierce Company, its sales affiliate in Mexico until 1911. Another subsidiary, the West Indian Oil Company, built refineries in Cuba and Puerto Rico that processed United States crude to avoid high tariffs on imported refined products. A similar effort to build a Standard plant at Rio de Janeiro was halted in 1896, when a tariff law favorable to the import of crude was changed abruptly. These refineries were important producers of kerosene and fuel oil, a product having a large market in the power plants, railroads, and steam ships of coal-short Latin America.[6] Meanwhile, the first direct sales branches had been set up at Buenos Aires and at Rio de Janeiro to satisfy the company's largest national markets in South America.[7] Standard Oil had little reason to move into its own production abroad as long as it had adequate supplies of United States crude oil and as long as no Latin American oil fields existed under the control of competitors. Neither of these conditions remained static, of course.

In fact, a number of independent British and American oilmen quite early were responding to the growing demand for kerosene and fuel oil in the Southern Hemisphere, and they soon began to gnaw at Standard Oil's sales dominance. These oil pioneers created Latin America's own petroleum industry to supply the railways and mines, to pave the streets of growing cities, and to illuminate the homes. The oil deposits of Peru's northern coast first were tapped in the 1870s by the American railroad builder Henry Meiggs, who had invested $150,000 in several wells and a refinery before his death. The initial wells sanded over during the War of the Pacific, and the Peruvian owner of the La Brea y Pariñas estate that Meiggs had leased actively sought a sale to British interests in 1889. Thereafter, the London and Pacific Petroleum Company, Ltd., under the management of English merchants, overcame chronic capital shortages in order to expand oil production to the modest range of 3,000 b/d by 1910, whereupon some twenty years without dividends motivated these owners to search for a buyer. By that time, production had begun at nearby Zorritos and Lagunitas, and a second British company at Lobitos was contributing another 1,038 b/d.[8] The Peruvian oil district was supplying oil markets along the west coast of South American from Valparaíso to Guayaquil through British mercantile houses.

Meanwhile, Mexican production was begun by competing American and British interests. Edward L. Doheny came to Mexico to find an oilfield that could supply fuel to the foreign-owned Mexican Central Railway. At that time, the company's locomotives were running on imported Alabama coal of uneven quality. Within a year of his arrival in 1900, Doheny had struck oil at Ebano; within five, his company was asphalting the streets of Mexican cities and was supplying fuel oil to American-owned railroads. By converting to domestic oil, the Mexican Central cut its fuel bill by 50 percent.[9]

British entrepreneurs, encouraged by a Mexican government wary of too much American domination, were not far behind. Sir Weetman Pearson (later Lord Cowdray) already had built Mexican railways, including a transisthmian line across Tehuantepec, and by 1910 he had opened up production not only on the Isthmus but also in Vera Cruz's famous Golden Lane. Pearson then reduced the market leadership of Waters-Pierce by cutting his domestic oil prices by one-third, allowing his company, El Aguila, to capture 40 percent of Mexican sales.[10] Standard Oil was encountering intensive competition in one of its oldest Latin American markets.

Somewhat the same phenomena also was taking shape in another large national market, except that in the Argentinian case, government well drillers had found oil deposits before any private firms had. The 1907 discovery of an oil field near Comodoro Rivadavia in Patagonia signaled a scramble for oil leases and concessions on the part of Argentinian and British companies. The government soon established a national reserve at Comodoro Rivadavia but allowed native speculators and foreign companies alike to take out claims to other land in the national territories of the south, while provincial governments likewise offered oil concessions to private entrepreneurs. By 1916, the government had established a refinery in Patagonia and was producing nearly 130,000 cubic meters of crude per year. Although private companies were adding another 8,000 cubic meters, Argentina's vibrant economy at the time consumed more petroleum than both private and public interests could produce domestically.[11] Paradoxically, Standard's growing import sales here were becoming increasingly insecure.

As competition slowly was encroaching on its foreign markets, Standard Oil's domestic situation changed abruptly. In 1911, the Supreme Court dissolved the giant Standard Oil group of companies and to separate the domestic refining and marketing subsidiaries from the producing companies. The New Jersey holding company in particular had retained its refining and marketing organization but was left with little vertical integration in oil production of its own.[12] Uncertain of its supplies of crude, Jersey now faced foreign competition with more weakness than its size suggested.

The years following dissolution witnessed the assumption of managerial control of the company by aggressive young executives who had a new concept of the international market and an outlook that would later be commonplace in the multinational corporation.

Walter Teagle for one had become familiar with overseas problems, especially the loss of markets to foreign producers like Royal Dutch–Shell, as a member of the Foreign Export Committee of the Standard group. Following dissolution, he became head of Jersey's Canadian affiliate, the Imperial Oil Company, Ltd., and was forced to seek production abroad because Canada had few oil fields of its own. He would continue this aggressive foreign activity when he was appointed president of Jersey Standard itself in 1917.[13] Two other executives associated with Imperial also were to loom large in Latin America. R. V. LeSueur, the chief counsel of Imperial, brought order to the oil fields of Peru, where he negotiated some early political problems. Eventually, he married a Peruvian from a distinguished family. In contrast to this austere lawyer was James W. Flanagan, a loquacious Texan with experience in Mexican railroads, the Cuban wars of liberation, and the lumber business. Flanagan utilized his gifts of charm and of business diplomacy to complete construction of the Colombian pipeline. Another proponent of foreign production at Jersey was Everett J. Sadler, who had managed Standard's first foreign producing company in Romania before World War I. Before becoming chief of the company's new Foreign Production Department, he supervised Jersey's first oil field in Mexico. Sadler was a tireless, even reckless, searcher for foreign concessions and opportunities. Together, these executives began aggressively to expand foreign production to supply established markets. Moreover, they considered acquisition of Latin American properties to be a rational method of meeting foreign competition while at the same time avoiding the limitations of United States antitrust laws.

The company's competition abroad was formidable. Its precursor in international production was the Royal Dutch–Shell group, having both Dutch and British ownership. Under the leadership of Sir Henri Deterding, Shell had pursued a policy of developing production close to foreign markets, and since the last decade of the nineteenth century he had been battling Standard Oil for markets in East Asia. Sir Henri effectively outmanuevered Jersey Standard in the Dutch East Indies, where colonial authorities resisted granting concessions to other than Dutch companies.[14] The same also might occur in British possessions, reasoned Jersey executives, where Shell and Anglo-Persian with government participation had the edge. Meanwhile, the invading German army in 1916 captured Jersey's operation in Romania as Sadler himself fled before the advance. Subsequently, the company lost some $11.5 million in the purchase of a half-interest in Russia's Baku oil fields at the very moment that the Bolsheviks seized the property.[15] These setbacks in Asia and Europe caused Jersey's executives to concentrate their attention closer to home—in Latin America.

Latin American production was logical from yet another standpoint. The markets there for petroleum products expanded greatly in the second decade of the twentieth century, several countries having converted to fuel oil during World War I as imports of British coal dwindled. Despite intense competition from other oilmen, Jersey Standard's net earnings from Latin American marketing increased from $1.5 million in 1912 to $7.9 million in 1918 and accounted for 43 percent of the company's total foreign sales.[16] Foreign marketing was growing fast even though the market share of Jersey was declining.

The war produced an additional boom to Jersey's plans for expansion abroad. Reduction of exploration in the United States during the war seriously had depleted known domestic reserves. The home government, confronted with a postwar coup of British and French diplomacy that seemingly excluded United States companies from the territories of the old Turkish Empire, was willing to promote the interests of North American oilmen abroad—even those representing Jersey Standard. The chief geologist of the United States Geological Survey in 1919 urged American oil companies to move aggressively into exploration of new sources of crude in the Middle East and South America. The secretary of state also instructed diplomatic representatives of the United States throughout the world to aid the country's oilmen abroad in obtaining foreign concessions and in competing effectively against rivals like Shell and Anglo-Persian.[17] But diplomatic support usually took the path of least resistance. As this chapter will show, American diplomats provided convenient liaison functions for oil companies in friendly environments like Venezuela while they demanded compromise and concessions from the companies in hostile environments like Mexico.

At any rate, Jersey Standard actually had begun to acquire Latin American producing properties before the announced diplomatic support. By the end of World War I, operations in Peru and Mexico were under way, and plans already were nearing completion for the purchase of a producing company in Colombia. Nonetheless, Teagle wasted little time writing to inform the State Department of Jersey Standard's activities. Having purchased Peruvian and Colombian properties for the Canadian affiliate, Teagle took pains to inform the State Department that these were essentially American operations.[18] However, the distinction often confused American legates both in Peru and in Colombia, who often questioned why they should aid a Canadian company that also claimed the diplomatic support of Canada and Great Britain.[19] Some American diplomats had difficulty understanding the new international organization of business. Likewise, Brazilian diplomats in the classic buffer states of Paraguay and Bolivia wrote home that a powerful new force was at work: the multinational oil company.

Standard Oil of New Jersey had the distinct disadvantage of beginning its quest for petroleum resources with a tarnished reputation. Literate Latin Americans, then as now, were avid readers of international affairs and had mastered well the muckraking accounts about the vicious business wars in the Pennsylvanian and mid-American oil fields. The supposed victor of those wars, Standard Oil, was held to embody all the evils of unfettered Anglo-American materialism, the perfect incarnation of José Enrique Rodó's Caliban.[20] Moreover, its large size suggested an acquisitive appetite that received justifiable punishment in the 1911 antitrust ruling. That Jersey Standard survived and grew afterwards was merely proof of the number of politicians it had on the payroll. The word "trust" had been admitted into the Spanish language, so common was its usage in the media, but it was by no means a synonym for *confianza*.

By contrast, the British oil companies headed by Sir Henri Deterding and by Lord Cowdray seemed models of respectability (note the titles of knighthood and peerage) and of corporate probity. For some reason, British businessmen appeared neither gauchely banal nor noveau riche, while the modest social backgrounds of many American entrepreneurs, including John D. Rockefeller, explained their reputedly unsavory business ethics and affronted many Latin Americans. The story of how Sir Henri had all but eliminated Standard Oil from the Dutch Empire, of course, had not made the world press and had not been the object of an antitrust suit.

In general, influential Latin Americans reacted to the entry of the foreign oilmen in several contradictory ways. The private national entrepreneurs formed an important group about which little is known. Before the arrival of the foreign companies, several had been oilmen in their own lands, working on primitive wells and stills or holding vast petroleum concessions they were unable to develop. A number of local businessmen grew wealthy by selling out to foreign interests.[21] Others sold oil leases and lands, collected royalties, worked as lawyers for the companies, and participated in domestic merchandizing of petroleum products.[22] Despite the fact that they provided little competition to the technological and financial resources of the foreigners, the national business communities in Latin America do not seem to have opposed foreign investments in petroleum. The principal nationalist opposition of the 1920s appears to have come from civil servants, politicians, and military officers.

Other Latin American political leaders, the nation builders, viewed the oil companies as engines for the modernization and strengthening of their countries. Porfirio Díaz of Mexico welcomed foreign investments of all kinds and viewed the development of national resources as the highest patriotic duty so that Mexico could avoid future foreign military invasions like those of the Mexican-American War and of the French intervention. Juan Vicente Gómez of Venezuela and Augusto Leguía of Peru were politicians of similar beliefs. The Peruvians were all too aware of their national humiliation during War of the Pacific, and the Venezuelans remembered the British fleet's enforcement of British control over disputed territory between Venezuela and Guiana. All three leaders justified their strong political methods as consistent not only with their country's political immaturity but also with their mission of building national power. For the most part, they and their supporters skillfully balanced the competitive instincts of the British and Americans, although opponents later accused them of *vende patria.*

Resistance to foreign economic influence was a sentiment that gained political currency in those countries already economically transformed by the combined efforts of the nation builders and foreign investment. Among sons of the elite and the emerging middle sectors, new motivations of pride and self-confidence brought conflicting goals to the political life of the region's more advanced economies. Mexico's revolutionists were intensely nationalistic and from 1917 onwards sought tougher restrictions and higher taxes on foreign interests. Brazilian leaders exhibited a persistent distrust of those few foreigners who were interested in oil exploration, while Argentinian politicians struggled to change the laissez-faire legacy of nineteenth-century liberalism. Occasionally, conflicting motivations set local and regional forces against each other. Within countries, the uneven pace of progress created dichotomies between urbanizing areas and the provincial countryside still yearning to taste the material fruits of foreign investment. In Vera Cruz, Mexico, and in Salta, Argentina, Jersey Standard operated in previously undeveloped areas where the local elites protected oilmen from the nationalists. Incipient nationalism, however, was present before the depression in all Latin American polities save Venezuela, but in fact it was strongest in those nations that had undergone an intensive process of nation building.[23]

The most difficult political environment for Jersey's production ventures was therefore found in those countries already having a long and intensive experience with foreign investment. British capital dominated Latin American money markets before World War I, and in 1913 its leading recipients were Argentina, Brazil, Mexico, and Chile; many of these acquaintances with British investors antedated 1880. Mexico's proximity to the expanding United States economy meant additional receipt of investments, and before 1917 North American businessmen here outnumbered their British counterparts. In terms of foreign-built railroads, mining and smelting operations, food processing plants, urban utilities, port operations, foreign trade, and urbanization, these countries were also the best markets for Standard Oil's petroleum sales. Their citizens already were familiar with the company. Jersey secured no production at all in Brazil and Chile, two large and vibrant national markets. Colombia and Peru, where Jersey Standard was able to operate despite some hostility, had been moderate recipients of foreign investment at the turn of the century. Likewise Venezuela, later the foreign oilman's most advantageous area of expansion, previously had been among the least attractive investment opportunities. British investment there in 1913 was a mere £2,004 sterling, just two percent of what British investors were spending in the Argentine republic.[24]

Those economic nationalists who resisted further foreign encroachment very early selected Standard Oil for special opposition. Standard was viewed as the epitome of the trust, intent upon monopolizing the world's reserves of crude oil in order to strangle small nations with price control. Rumor that Standard Oil financed the rebellion of Francisco I. Madero against the Mexican government of Porfirio Díaz (who incidently was wary of Standard Oil's reputation) had unsettled many politicians in Latin America, and through a short flight of imagination many implicated the trusts in the United States military occupation of Nicaragua. Ecuadorian legislators had had little direct experience with Jersey's operations in an area in which sales were not yet extensive, yet they were motivated to resist the company's efforts at exploration in favor of British oilmen. Standard was depicted as a boa constrictor, "the hissing of which is heard on the shores of Zambesi and the Lualaba, which swallows up little birds like Nicaragua and chokes powerful antelopes like Mexico."[25] A more typical case is Argentina, where Jersey Standard did have a refining and marketing apparatus but where, unlike some British interests before the war, it had not yet attempted to secure even a single oil lease. Nevertheless, in 1913, the chief of Argentina's petroleum board felt compelled to state: "Throughout the world, Standard Oil acts like a band of cruel, usurious pirates, headed by an ex-clerk, who began by carrying thousands of families among his own countrymen to ruin. Like an octopus, [Standard Oil] has extended its tentacles everywhere, accumulating colossal fortunes of millions of pesos on the basis of human blood and tears."[26]

Such notoriety, out of all proportion to the company's activities in Latin America, drew Standard Oil Company (New Jersey) into political turmoil nearly everywhere it entered as producer. Not that Jersey executives were unprepared for or even baffled by early opposition. Their experience among cut-throat rivals in United States oil fields was already extensive. But the North American business culture of the day dictated that Jersey's response to the economic nationalists as well as to competitors was secrecy, a tactic that eased entry but thereafter poisoned the company's relations with host governments.

Formation of the IPC in Peru

The pattern of Jersey's entry into production was first established in Peru, where the near-monopoly status of this infamous world oil conglomerate embarrassed the government. In truth, the company had wanted to begin in Mexico, not Peru. Jersey's attempts in 1912 to purchase El Aguila, largest foreign producer in Mexico, had failed, largely because the price was high and Jersey preferred to continue purchasing crude from a number of independent Mexican producers.[27] Company executives decided to look for opportunities elsewhere.

Expansion into Peru was not without its detractors within Jersey headquarters, one of whom was its president in New York, John Archbold. Presented with the opportunity in 1910 to purchase the oil fields of the oldest producing company in Latin America, the London and Pacific Petroleum Company, Ltd., Archbold instead recommended arranging for the long-term purchase of Peruvian crude. In 1911, Standard formed the West Coast Oil Fuel Company, Ltd., with minority ownership shared by the London mercantile houses that controlled Peru's oil industry. The new organization facilitated the exchange of Peruvian crude suited to California's gasoline markets for the heavy California crude oil ideal in South American fuel sales. However, Archbold agreed in 1913 to a field survey of the properties in Peru at the suggestion of Walter Teagle. The preliminary report emphasized the inability of the British amateurs to develop properly the large potential of the properties. Obviously, the London and Pacific, a trading company, lacked both the personnel and the capital to establish a thriving oil operation on Peru's barren north coast. The 1913 report speculated that the oil fields on the large La Brea estate could be worth four times the option price if expertly developed.[28]

Once New York's resistance was overcome, the Canadian affiliate, Imperial Oil, Ltd., whose president at the time was Teagle, purchased the British company in Peru. Thus, Jersey Standard surmounted the Englishmen's reluctance to sell to an American company and avoided any legal complications arising from United States antitrust laws. The Canadian affiliate created the International Petroleum Company (IPC) with capital stock valued at £4 million.[29] Control of administration shifted to Toronto, which conferred with New York in the marketing of oil products. The only Peruvian oil interests outside of IPC's control were a native company at Zorritos and a British company at Lobitos, both of which were to retain a separate existence from the IPC while at the same time selling it crude oil.

The new management undertook steps immediately to improve Peru's oldest oil field. In 1916, Teagle dispatched to Peru two of the company's best engineers, one a refiner and the other a producing man, along with North American drillers, while experienced company geologists made extensive surveys of the La Brea estate.[30] Production increased until 1916, when Peruvian laborers struck for a 50 percent wage boost. The government, fearful of losing essential revenues from oil exports, dispatched soldiers to protect company property. The strike at Talara ended peacefully without the company yielding to wage demands, but the labor unrest at nearby Lobitos led to a clash between soldiers and strikers in which several workers died.[31]

Output of crude began to justify early confidence in the prospects of the Peruvian oil fields. By 1921, International had drilled more than 487 new wells and increased production on the La Brea estate to some 7,741 b/d of crude. Its expanded refinery at Talara processed most of the La Brea crude, an additional 2,000 b/d that it purchased from the independent Lobitos company, and quantities of heavier California and Mexican crude for fuel oil.[32] Obviously, Jersey Standard was utilizing its resources and expertise to bring new life to the Peruvian oil fields, whose production increases would have been impossible without the company's marketing flexibility.

Selling Peru's petroleum proved an intricate process, combining control from Toronto and New York and the use of traditional merchant outlets. The IPC maintained its marketing connection with British merchant houses on the Pacific coast of South America. London's house of Balfour, Williamson already maintained well-established general agents for the sale of fuel oil to West Coast railroads, Chilean nitrate companies, and Peruvian mining concerns. Meantime, the IPC controlled the sale of its illuminants in Peru, establishing in 1919 Lima's first bulk distribution plant for kerosene. For the moment, Peru's petroleum was unsuitable for a South American market in which motor cars and lorries were a rarity, so the light crudes went to refineries in Vancouver and New Jersey, while a Jersey Standard marketing subsidiary, the West India Oil Company, sold Peruvian petroleum products elsewhere in the more developed petroleum markets of Latin America.[33] Expansion of the IPC in northern Peru had enhanced the country's early ranking as the number two oil producer in Latin America.

Jersey Standard encountered political resistence that threatened its operations in Peru altogether. It had concluded its Peruvian purchase in an atmosphere of secrecy that was to be characteristic of its entry into many Latin American nations. The transfer of control of the London & Pacific had escaped notice not only of the British public for a time but also of Peruvian newspapermen and politicians. It is conceivable, although not proven conclusively, that the Peruvian government's embarrassment as mounting evidence indicated momentous changes in the nation's oil fields impelled it ultimately to decree higher taxes on this notoriously wealthy foreign company. The host government claimed an error in the original 1888 measurement of the property and decreed that the mining tax on the La Brea y Parinas estate be raised from £30 per year to £120,000.[34] Thus began the government-industry controversy that was to haunt IPC's operations in Peru even after its settlement seven years later.

The government's infringement on IPC's contractual rights moved company officials to respond according to a pattern that was to typify their behavior during later controversies elsewhere in Latin America. Imperial's executives from Toronto assumed control of the IPC's legal defense, with Walter Teagle adding his counsel from New York once he became president of Jersey Standard in November 1917. R. V. LeSueur of Imperial's legal staff traveled to Peru and hired a leading Peruvian jurist to help him prepare the IPC case. Company officials made known their objections to the new taxes that, they said, would render the oil fields commercially unworkable, possibly forcing the IPC to abandon the property completely. LeSueur, in the meantime, prepared a legal defense based upon the premise that higher taxation constituted an unconstitutional violation of the private contract that Imperial had purchased in good faith from the London and Pacific.[35]

Relinquishment of private property rights may have been what the Peruvian officials had in mind, for state control of subsoil deposits was an idea beginning to regain currency throughout Latin America. The concept of state ownership of subsoil wealth hearkens to Latin America's colonial legacy, which retained for the monarchy the rights to all mining activities in the realm. Politicians influenced by nineteenth-century liberalism, on the other hand, had eliminated many of the crown's mining regulations as an unwanted inheritance of Latin America's colonial past. Numerous mining laws enacted in the previous century in fact had adopted the

English common-law concept of the underground mines as the private property of the surface owner. Nation-building liberals had enacted such laws in Mexico, Brazil, and Argentina, just as the Peruvians had allowed individual cases of private mining properties. For their inspiration, however, twentieth-century nationalists looked not to the laissez-faire liberalism that so inspired the nation builders but to the Spanish legal heritage of state control.

Cognizant of this Hispanic tradition, Jersey Standard attorneys preferred to negotiate a settlement with the government—a decision that opened for public debate the very status of the company in Peru. Teagle and other Imperial executives traveled to Lima in 1916 to confer with Peruvian officials. International's executives made an interim compromise with President Manuel Pardo but the Peruvian Congress refused to ratify it. Meanwhile, the government's chronic need of funds complicated the issue. When Peru requested a loan of $3 million from several private banks in New York, the bankers required settlement of the petroleum issue first, so that export and mining taxes could be fixed to secure repayment of the loan.[36]

The problem was confounded by the fact that Peruvian political life was disintegrating rapidly as President Pardo lost support from influential members of the *república aristocrática*, the coalition of elites that successfully had eliminated the military from politics and had preserved governmental stability for three decades. Congressmen were uneasy about the possibility of President Pardo imposing his handpicked successor in the coming year's elections and were unable to act on the petroleum issue.[37] The company in 1918 therefore resolved to settle the stalemate according to its own timetable. A shipping shortage and company annoyance at the political impasse provided the pretext. LeSueur decided to curtail IPC's operations, laying off six hundred workmen at its oil fields and shutting down the refinery. Then Canada requisitioned for wartime use two tankers that had been carrying petroleum from Talara to the port of Callao.[38]

Lima was faced with an oil shortage, and Peruvian editors and congressmen were furious. *El Comercio* advised the government to form a native company to exploit the oil resources. International soon relented to an order by President Pardo for return of the tankers as a precondition for resolution of the impasse.[39] Congress then authorized the executive to enter into an agreement that would be submitted to an international arbitor, as had been suggested by the IPC.[40]

Despite IPC's small tactical gain, final resolution of the controversy awaited the accession of a new president. Augusto B. Leguía assumed office just as a postwar depression struck the Peruvian economy, for a time depriving the government of the money to pay its growing staff of bureaucrats. There is also evidence that not all the oilmen approved of the new executive. More than once, the United States ambassador felt compelled to summon IPC's Lima representative to warn him against public statements predicting Leguía's downfall.[41] However, given his extensive business background, the new president turned out to be a nation builder and not an economic nationalist. In an interview with Peru's English language newspaper in 1919, Leguía stated that "during my administration foreign capital will be given facilities and opportunities for the development of Peruvian resources which have never heretofore been accorded and which may never be accorded again."[42] The president, secure in his new political position, nonetheless drove a hard bargain, requesting that the IPC make a gift of one million dollars to the government as a precondition to settlement. Walter Teagle in New York gave his consent.[43] Thereupon, IPC's executives and Leguía's advisors worked quickly toward a resolution of the controversy.

In a compromise signed in 1922, the government conceded that La Brea's subsoil was to be considered private property. In return, the IPC agreed to higher taxation on the La Brea estate but secured the right to abandon any part of the property to avoid paying taxes on unpromising land. In addition, the tax rates were not to be changed by subsequent petroleum laws until 1972. Great Britain arranged for the ratification of the agreement in Geneva, where a panel consisting of representatives from Canada, Peru, and Switzerland then handed down the award as agreed upon in advance.[44] It became known as the 1922 Accord.

The year 1922, therefore, forms the watershed of Jersey Standard's operations in Peru. During the seven-year controversy, net production at La Brea fluctuated between 5,000 and 7,700 b/d. Solution of the outstanding title and tax issues encouraged the company to undertake an expansion program that yielded more than 29,000 b/d by 1928 (see table 2). International's fixed assets nearly doubled, and the number of Peruvian employees increased from 2,552 in 1922 to 6,143 in 1927, with a foreign staff of 340 persons.[45]

Table 2: Daily Average Production of Crude Oil by Jersey
Standard Affiliates (in thousands of 42-gallon barrels)

	Peru	Mexico	Colombia	Venezuela	Bolivia	Argentina
1915	5.0					
1916	5.1					
1917	5.0					
1918	5.0	2.6				
1919	5.1	21.2				
1920	5.4	38.6				
1921	7.7	36.5	0.2			
1922	12.0	13.2	0.7	0.0		
1923	12.7	58.7	0.8	0.1		
1924	17.7	50.2	0.5	0.0		
1925	20.1	45.6	1.9	0.0		
1926	23.7	19.1	15.6	0.0	0.1	0.7
1927	21.3	9.8	36.7	0.1	0.1	0.8
1928	26.0	6.4	48.2	14.2	0.0	1.2
1929	29.6	5.0	49.3	18.6	0.2	1.9
1930	26.8	4.5	49.1	17.1	0.1	2.1
1931	21.0	4.1	43.8	20.4	0.1	3.3
1932	20.9	13.5	39.3	82.4	0.1	4.7
1933	30.7	21.3	31.6	123.9	0.3	6.4
1934	38.7	27.3	41.6	160.4	0.4	6.6
1935	40.4	18.6	32.3	181.1	0.4	5.8
1936	41.3	11.9	45.3	198.6	0.3	5.2
1937	40.3	16.1	49.5	236.6	0.0	5.2
1938	36.0	2.0	52.5	233.3	—	4.9
1939	29.6	—	54.5	273.0	—	4.6

Sources: George Sweet Gibb and Evelyn H. Knowlton, *The Resurgent Years,
1911–1927* (New York, 1956), pp. 676–77; and Henrietta M. Larson, Evelyn
H. Knowlton, and Charles S. Popple, *New Horizons, 1927–1950* (New York,
1971), p. 115.

Throughout its foreign operations, Jersey's policy was to employ as
many native workers as possible, although the presence of a large
number of experienced foreigners initially was necessary to expand
the operations and to provide on-the-job training to the Peruvians.
To the IPC, as to other multinational corporations since, it simply
made good political sense to train Peruvian drillers, mechanics, and

office workers, while the gradual reduction of the foreign staff helped lower expenses. The year 1922 illustrates another characteristic of Jersey Standard's operations in Latin America: political uncertainty tended to dampen investment in oil production while the resolution of conflict encouraged investment and expanded production (see table 2).

The arbitration agreement, nevertheless, permanently poisoned relations between the IPC and Peru. Although the company subsequently explored for new oil fields in Peru's Oriente region and in the Sechura Desert without success, all of the Jersey Standard's activities remained at its La Brea site until 1968, when the government, using the "illegality" of the 1922 Accord as one of its justifications, nationalized the assets of the International Petroleum Company.[46] Its early political problems in Peru served as an introduction to Latin American politics and to the perils of economic nationalism. Governments elsewhere also were revising the terms under which the foreign companies were operating.

Mexican Operations

In Mexico, oil development and revolution occurred simultaneously in the second decade of the twentieth century, and they clashed for two reasons: foreigners controlled the growing industry, and economic nationalism became an integral part of revolutionary ideology. As in Peru, foreign oilmen considered the subsoil wealth to be the private property of the landowner and his leasee, while the revolutionary government went beyond the Peruvian case in seeking to gain control of the nation's oil resources by resurrecting the Spanish mining tradition. All comprise between politicians and the companies proved temporary, but in fact Jersey's political position in Mexico was badly weakened by 1928. Much of Jersey's problems here resulted from the wasteful and uncontrolled competition in oil districts to which the company arrived late.

Following its failure to purchase a producing company in 1912, Jersey had to be content with acquiring crude from other producers. Its skimming plant at Tampico separated the heavy crude into fractions suitable for further conversion to asphalt and fuel oil in its United States refineries.[47] At the same time, nineteen companies were operating in the Golden Lane of the state of Vera Cruz. All held oil leases secured during the presidency of Porfirio Díaz, whose petroleum laws had overturned Spanish legal tradition concerning sovereign ownership of subsoil wealth. As the 1884 law defined the subsoil minerals as the personal property of the owner of the surface land, prerevolutionary oil leases were private contracts to which the government was to have no claim.[48] Jersey Standard could not hope to secure such beneficial concessions under the revolutionary governments.

Within two months of the promulgation of the Mexican Constitution of 1917, Jersey purchased the Compañía Transcontinental de Petróleo, S.A., as its Mexican producing affiliate. The sales price of $2,475,000 (U.S. gold) included a Mexican charter, valuable Díaz-era leases, and a small amount of existing production.[49] Transcontinental had but one producing well, but some of its concessions included properties in the southern fields of Vera Cruz, where a valuable light oil had been discovered. Everett J. Sadler was convinced that the new affiliate, in a few years, would be able to deliver some 100,000 b/d of Mexican crude.[50] His initial report from Mexico was unduly optimistic. So eager was Sadler to get Jersey Standard involved in the Mexican bonanza that he mentioned not a word of Article 27 of the new constitution, which nationalized the subsoil minerals, or of the unfortunately chaotic conditions in the fields. The reason: Mexico was to be an integral part of the company's new strategy to meet and beat the competition abroad.

As World War I came to a close, Jersey's competitors also were strengthening themselves in foreign production. Its chief rival, Royal Dutch–Shell already had extensive production in the new fields of Venezuela, having struck several gushers in the Maracaibo Basin. Then in 1919, the Shell group purchased the largest Mexican company, El Aguila of Lord Cowdray, which had a daily production of nearly 120,000 b/d and a suspected potential of 700,000 b/d.[51] Once again Shell had outmaneuvered Jersey Standard, this time in its own back yard!

Standard Oil Company (New Jersey) responded by launching a vigorous policy of capital formation to expand its operations both at home and abroad. For the first time since the dissolution, Jersey went into debt. The board of directors in 1919 authorized an increase in the preferred stock for two years following the war, and in 1926 they issued 5 percent debentures to pay for redemption of the preferred stock.[52] By 1927, the common stock of the international company had grown to a total of $750 million. Its assets dwarfed the national budgets of countries that the company was entering, a fact that seldom escaped the nationalist politicians and confirmed their fears about the power of the trusts.

Jersey's first priority in 1919 was the expansion of production in Mexico at a time when Mexican production had grown to the second-largest in the world, behind that of the United States. The dollar investment in the Transcontinental affiliate grew nearly to $33 million in 1922. What new leases it was able to acquire afforded the company future interests in northern Mexico and on the Isthmus of Tehuantepec.[53] Production burgeoned as several of the new Pánuco wells proved to be gushers and the first two producing wells in the southern fields yielded a valuable cache of light crude. Transcontinental's production rose from 2,552 b/d in 1918 to a peak of 58,713 b/d in 1923.[54] Moreover, the subsidiary purchased additional crude from Mexican companies for its topping plant and for a newly purchased refinery near Tampico. Transcontinental easily recouped the initial investment of the parent company, repaying all the loans to Jersey Standard in 1924 and in the next three years, passing on a total of $36.5 million in stock dividends.[55]

Nevertheless, the Mexican venture rapidly was losing its vitality because of the prolifically ruinous competition in the Vera Cruz fields. Oil prospecting here had become frenzied after 1910, as companies scrambled to secure leases from the numerous farmers who owned small parcels of land over the principal oil pools. In the Pánuco field, no less than eight different oil companies were operating within an area of several square kilometers.[56] Such conditions gave rise to a furious offset campaign, duplicating the worst features of oil competition in the Pennsylvania, Texas, and Oklahoma oil fields, in which each company sought to drill more wells and to pump more crude from the common reservoir before the neighboring operator. Wells caught fire, consuming valuable gas and crude, while gas escaped between fissures that broke through the ground between the closely drilled wells. Individual American workers were encouraged to finance their own shallow wells in a speculative bid to become wealthy overnight, a practice that Transcontinental by 1920 had ended on its leaseholds.[57] Underground gas pressures that previously had assured good flow rates at the wellheads soon began to dissipate, and the wells started producing saltwater instead of crude petroleum.

An effort to cooperate with other oil companies in reducing the flow of the wells accomplished little. Transcontinental abandoned its source of light crude in the southern fields altogether.[58] Jersey's prospectors meanwhile were traveling throughout Mexico in search of new petroleum resources. They surveyed the Gulf Coast from Laredo, Texas, to the Isthmus of Tehuantepec, and scouts ventured into the north and into Oaxaca and Guerrero in the south.[59] Having acquired some options in these districts, Jersey Standard executives nonetheless made the decision not to pursue them. Although no one seemed pessimistic about the prospects of new oil discoveries in Mexico, the unrelenting political pressure of the Mexican government had the effect of containing the foreign oilmen's operations in Mexico.

Until 1920, the operations of Transcontinental in Mexico had been relatively free from revolutionary turmoil. Like other companies in the area, the subsidiary had made forced loans totaling some $59,000 to the military chief of Vera Cruz to protect the installations and workers.[60] During the de la Huerta rebellion of 1924, mounted raiders broke into the camps and halted operations. Then a strike suspended activities until company executives agreed to raise the wages and improve working conditions, although a union demand that the workers share in the management of the company was rejected.[61] Similar issues would reappear in the labor dispute that led ultimately to nationalization in 1938.

The principal threats to Transcontinental emanated neither from raids nor from labor strikes but from the revolutionary government in a mounting controversy over Mexico's Constitution of 1917. Even though Article 27 stipulated that all disputes were to be settled in Mexican courts, the foreigners had counted on the diplomatic protection of their home government because their most valuable concessions and oil leases antedated the constitution. For example, Jersey had purchased leases made during the Díaz era, when hydrocarbons were considered private property, a condition that conflicted with Article 27, which stated that all subsoil minerals belonged to the nation. The foreign investors, therefore, based their position on another part of the constitution stating that no law should be given retroactive effect.

At the moment that the southern fields were being exhausted of light crude, President Álvaro Obregón decreed an increase in the oil export tax. Jersey's response was that the new tax was confiscatory. Since the tax would raise the cost of the poor-grade crude upon which the companies in Mexico had to rely now that the light crude was running down, North American oilmen claimed that products refined from heavy Mexican oil could be undersold by superior United States midcontinental crudes.[62] Jersey officials argued that the poorer Mexican crude needed to remain inexpensive or lose its markets. Obregón agreed to a smaller increase in export taxes, but only until the end of the year, thus postponing rather than resolving the issue. Although extensions of the postponement were granted, lack of a definitive resolution merely added to the oilmen's insecurity.

The proposed petroleum laws of 1922 further threatened the foreign investors in Mexico. What troubled Jersey most was whether Article 27 would be interpreted to give the government the subsoil rights on lands owned by Mexicans before 1917, because Transcontinental had acquired more leases since 1917 than had most of its competitors.[63] Therefore, Jersey was more vulnerable than were the other foreign oil companies for having entered late. Another compromise failed to offer the Jersey company much in the way of assurance. At the Bucareli Conference of 1924, the Obregón government conceded that it would not nationalize oil resources on lands that had been developed before 1917.[64] The problem for Jersey was that its pre-1917 fields were nearing exhaustion and that Transcontinental had wanted to expand into lands acquired since 1917. Those expansion plans remained on the shelves.

By the time that Plutarco Calles had become president in 1924, the United States Department of State had taken over from the companies the responsibility of negotiating petroleum issues. By then, the mutual suspicion between the Mexican government and the foreign companies was absolute. Just the same, Calles was able to split the united front of the oilmen when he proposed a petroleum law to take effect in 1926 that would have struck Transcontinental very hard. The Calles oil legislation nationalized the subsoil of lands owned privately before 1917 but leased after that time.[65] Jersey's competitors, the Doheny and Shell companies, had reason to be satisfied because both had an abundance of pre-1917 leases, and their profunctory protests concerned principally a concurrent increase in taxes. At this juncture, Dwight Morrow arrived in Mexico as the new United States ambassador.

Internal debates over the Mexican oil problem were occurring both in the State Department and in Jersey headquarters. As evidenced by a working paper prepared within the department, diplomatic analysts decided that they were unwilling to support the legal arguments of the oil companies. The 74-page study concluded that, in view of colonial Spanish mining laws, the companies acquiring property before 1917 must have been aware that the original theory of sovereign ownership of mineral deposits might be placed back into force at any moment.[66] Meanwhile, Jersey officials in New York were rejecting a compromise position that might have preserved some of its assets in Mexico, albeit with high taxes on its post-1917 leases. Instead, they opted for a continental perspective. "Taking a long look ahead . . . and considering our growing interest abroad," concluded a Jersey legal counselor, "it seems to me that it is vital to us to stand for the sanctity of property and property rights."[67] The State Department, however, undertook the negotiations.

The Calles-Morrow agreement of 1928 resulted in two compromises, both of which severely damaged Jersey's legal status in Mexico. First, land acquired before 1917 was to be exempt from nationalization. Second, Morrow conceded that further questions effecting the operations of the foreign oil companies in Mexico were to be settled in the local courts.[68] When the contents of the Calles-Morrow agreement became known, Jersey Standard representatives wrote a 16-page letter of protest to the State Department, the tone of which indicated clearly that the oilmen did not consider the new policy a protection of American business interests in Mexico.[69] In view of the increased cost of post-1917 leases and its relegation to rapidly exhausting fields, Jersey interpreted the agreement as a defeat.

From the perspective of their headquarters in New York, the executives of Standard Oil of New Jersey perceived other opportunities in Latin America. One Jersey man concluded, "I believe in principle that oil cannot be produced profitably from [the post-1917 leaseholds] in Northern Mexico under the present tax, royalty and other obligations that the Mexican Government imposes; therefore, I would not spend good money after bad."[70] Operations in Mexico were maintained in the case that the political climate would improve. Meanwhile, Jersey's "good money" was being sent into Colombia, into Argentina, and increasingly into Venezuela.

Tropical Venture along the Magdalena

In Colombia, Jersey Standard was to follow a course of action much
like the one it was undertaking in Peru and Mexico. It was able to
buy and improve production on an existing concession but discovered that the public debate between the nation builders and the
nationalists soon placed the company under public scrutiny that
for a time threatened its Colombian operations. In 1919, Jersey purchased the De Mares concession in the Magdalena River valley at
Barrancabermeja. With the property went Tropical Oil, a Pitts-
burgh-based company that already had sunk three wildcat wells of
which only one produced.[71] Jersey executives had visited the De
Mares and other oil properties, surveyed political conditions in
Bogotá, and reported to the New York office. Their memoranda resulted in a decision to purchase Tropical for the Canadian affiliate,
Imperial Oil, Ltd., a strategy calculated not to antagonize Colombian sensitivity about the active role the United States had taken in
1903 to secure Panamanian independence.[72]

The Colombian venture was risky. The De Mares oil fields were
unproved and foreign oilmen considered the nation's new petroleum law of 1919 to be very restrictive, requiring excessive taxes
and royalties, limiting each concession to 30 years, and providing
for state takeover at the end of the term. The last stipulation was an
innovation favorable to the national government and one that
oilmen had not yet encountered on entry into a Latin American
country. Even so, the government was reticent about granting petroleum contracts under the law. Although it had received numerous applications, the government had made no grants at all within
the first two years of the law.[73] But the De Mares concession, dating
from 1905, contained more favorable conditions than could be acquired on newly granted concessions under the new petroleum law.
For once Jersey Standard's position· was stronger than that of competitors who entered Colombia later.

Nevertheless, the government refused to sanction the transfer of
the De Mares concession until the multinational company agreed to
additional terms. Tropical was required to build a refinery in Colombia, pay 10 percent production taxes, and give up the concession within thirty years of the beginning of production. The state
relinquished its right of ownership to the petroleum mines but stipulated that Tropical's operations were to be subject to the jurisdic-

tion of the national courts. In addition, at least a quarter of the supervisory personnel were to be Colombians.[74] Obviously, Colombia's politicians had learned about the contractual ambiguities then
plaguing Mexico's oil industry and resolved to avoid at the outset
such misunderstanding.

Development of production at De Mares was the most difficult
that Jersey Standard had encountered anywhere up to that time.
Located 560 kilometers up the Magdalena River, the concession
was covered with dense tropical growth and lacked roads and living accommodations. Tropical's first five years were employed
chiefly in building an infrastructure for its subsequent operations.
Varied projects were carried out simultaneously: constructing living quarters, utilities, terminal facilities, a railway, and a refinery at
Barrancabermeja; opening trails and roads through the jungle; acquiring river transport; studying the geology of the concession; and
surveying, wildcatting, and clearing well sites in the jungle. Even
farms had to be established nearby to provide cattle, hogs, cassava,
and bananas for the employees.[75] The refinery to serve the domestic
Colombian market was completed within a year at a cost of $1 mil-
lion. Although connected to the nearby oil fields by pipeline, it
initially processed some 1400 b/d mostly of imported crude that
was transported upriver in barges. Imported crude was not needed
for long. By the end of 1926, workmen had completed 141 wells,
only 13 of which proved to be dry holes. The producing department
estimated that the potential resources of De Mares equalled approximately 174 million barrels, sufficient to supply an export pipeline
for ten years at the rate of 50,000 b/d.[76]

Tropical's first obligation in marketing, however, was to supply
Colombia's domestic needs. Jersey's oil importing subsidiary, the
West India Oil Co., withdrew from Colombia, and thereafter Tropical sold its own products through local Colombian merchants. Although the company maintained a daily credit balance that averaged more than $300,000, its losses from defaults over a five-year
period, remarkably, were less than $300. Domestic sales rose from
73,543 barrels in 1922 to 987,853 in 1927, in large measure because
Tropical's prices undercut those of imported products. In Cartagena, where imported kerosene had been selling at $6.30 per 10-
gallon case, Tropical's price was $2.95, and low-priced Tropical
fuel oil soon replaced wood and imported coal in Colombia's river
and coastal vessels.[77] The import substitution process and the con-

version of the economy to petroleum occurred simultaneously, followed by the export of Colombian oil.

From the beginning Tropical encouraged the employment of as many native Colombians as possible. Contractors brought in workers from highland Departments of Santander and Antioquía. In 1926, the company was employing 1,850 men in the producing fields and 429 at the Barrancabermeja refinery; only 140 were foreigners.[78] The first years witnessed a high turnover rate among native workers accustomed both to seasonal agriculture and to temperate climates. To stabilize the labor force, Tropical managers favored employing the *antioqueño* as a more disciplined worker. The high incidence of tropical disease (146 cases of malaria were reported in July 1921) occasioned extensive work by the medical staffs both of Jersey Standard and of the Canadian holding company. By 1926, the base hospital at Barrancabermeja and the smaller hospital and field dispensary in the oil fields were staffed by seven doctors, one dentist, six nurses, and one hundred six other men and women.[79] Improvement in the camps and in the medical facilities eventually promoted a stable labor force.

Development of the De Mares concession incurred heavy investments. By the end of 1927, Tropical had assets and liabilities of more than $92 million, and after six years of work, the investment finally was beginning to pay off. Net income for 1927, the first year of full commercial production and of export, reached nearly $7 million.[80] Jersey Standard's efforts in the Magdalena River valley alone had propelled Colombia past Peru as the number-three petroleum producer in Latin America, behind Mexico and Venezuela.

Other petroleum companies had been awaiting Tropical's progress, and its success by 1926 encouraged extensive foreign interest in further exploration. Tropical itself was seeking addition concessions, despite the legal department's opinion that the Colombian petroleum law was among the most unattractive in South America, and the subsidiary continued extensive geological surveys in Colombia. The biggest competitors seemed to be Gulf Oil, Standard Oil of Indiana, Standard Oil of California, and Anglo-Persian.[81] A Colombian oil boom appeared about to begin.

What began instead was a government crackdown on the foreign oil companies, led by a new administration that was more nationalistic in outlook. Political controversy had been building even as Tropical was bringing in its production, and public opinion began to over-estimate the nation's petroleum resources. The underlying cause for the political opposition was Colombia's rising national debt that culminated in a temporary default of bond repayments in 1927. Some politicians and editors thought that Standard Oil (New Jersey) ought to be paying more for the right to exploit the country's "unlimited" oil resources.[82] Tropical's success made it a natural target of political opposition, but an unfortunate subterfuge perpetrated by Jersey Standard only confirmed Colombians' suspicion of the foreign oilman's unsavory, backstage political power.

Soon after Teagle had purchased Tropical in 1919, he decided to form an "independent" pipeline company. The Tropical contract already provided the authority for Jersey to build a pipeline from De Mares to the port of Cartagena, yet it was not needed until production began at the oil fields, still some years ahead. Therefore, Teagle conceived a plan for a separate pipeline concession that did not have to be turned over to the government in thirty years and that did not have to carry government royalty oil at fixed rates. He wanted a new pipeline contract more favorable than the one Jersey already had.[83] To acquire it, Teagle contacted his sometimes agent, a flamboyant American businessman named James W. Flanagan, a man whose connections with Jersey Standard were kept a secret.

Flanagan introduced himself in Bogotá's society as a British citizen and as general manager of a Canadian company, the Andean National Corporation, independent of any oil interest. Without much trouble, he obtained the necessary properties and leases required to bring the pipeline from Barrancabermeja to Cartagena, although gaining permission from the government to construct the pipeline proved more difficult. Flanagan established contacts in Bogotá with the British minister, the papal nuncio, the archbishop, and influential Colombian families. Correspondence between the two men indicates that Teagle did not grant his agent authority to use bribery—much to Flanagan's chagrin.[84] At one point, Flanagan traveled back to Washington to work unofficially for ratification of the United States–Colombian treaty that was to settle the Panamanian affair. When he managed to introduce the Colombian ambassador to Senator Albert B. Fall of the Senate Foreign Relations Committee, neither man knew of Flanagan's connection with Jersey Standard.[85] The treaty passed in 1921.

Back in Bogotá, Flanagan pressed anew for government permission to build the pipeline. Government ministers had agreed to a draft of the contract when news broke that Andean was connected to Standard Oil. Bogotá's newspapers, already filled with the details of the Teapot Dome Scandal, criticized the foreign oil interests and those politicians cooperating with them.[86] Flanagan's tactics, condoned by Teagle, had backfired in the worst way, and Tropical's operations came under increasing scrutiny.

Tropical had been able all along to construct the pipeline according to the terms of its existing contract. Using the Andean leaseholds secured by Flanagan, pipeline engineers began work at Barrancabermeja in 1925, reaching Cartagena the following year. At 538 kilometers, the Magdalena pipeline was the longest ever built in a tropical environment. Supervised by men who had been constructing pipelines in West Virginia and Pennsylvania, the crews laid specially ordered screw-type, 10-inch pipe manufactured in Pittsburgh. Cranes, barges, paddle wheel riverboats, Caterpillar tractors, and heavy wagons were imported to transport men and pipe to the work sites. More than four thousand Colombian workers and four hundred foreign technicians laid the line through the jungles and swamps of the Magdalena valley.[87] The pipeline was an engineering marvel, but political opposition was growing.

The new administration that assumed office in 1926 was committed to greater regulation of foreign businessmen and oil companies. The minister of industries cancelled the Barco concession of Gulf Oil, and a congressional committee recommended the invalidation of Tropical's De Mares concession. Proposed oil legislation offered higher taxes, more state control of pipelines, and state financial participation in oil development, although a congressional stalemate between the nationalists and the nation builders blocked passage of all oil legislation. During the following decade of rancourous debate over oil issues, the other oil companies abandoned their concessions in Colombia, leaving Tropical with its favorable 1919 De Mares contract as the monopoly oil company in Colombia. Government monitoring thereafter was intense, domestic petroleum prices were fixed, and few additional oil concessions were forthcoming.[88] In the meantime, Jersey Standard was facing a challenge of a more direct nature in Argentina, where its subsidiary competed not only against rival foreign companies and against nationalist politicians but also against a state petroleum corporation.

Pioneering in the Southern Cone

Among the last of the private oil companies to arrive in Argentine production, Jersey Standard nonetheless became the focus of an intense internal political struggle between the nationalists in the federal government and the nation-building politicians who controlled one of the important oil-producing provinces. Jersey's entry into Argentina and Bolivia occurred in the 1920s because of the rapid growth of markets for its oil products in the Southern Cone of South America. Already a leading market for imported energy supplies before World War I, Argentina doubled its consumption of petroleum products between 1921 and 1927, with gasoline sales growing the fastest.[89] Moreover, a high tariff on imported oil stimulated development of domestic sources of crude for Jersey's refinery at Campana, north of Buenos Aires. Incorporated in 1921, Standard Oil of Argentina began to obtain concessions and drilling rights in Patagonia, where government drillers already had been producing oil since 1907, and in the western provinces of Mendoza and Salta.

The Argentinian affiliate undertook extensive exploration, but results in the south were not encouraging. Several dry holes in the region of Comodoro Rivadavia induced the company to abandon all but 10,000 acres.[90] By the end of 1927, Standard Argentina had expended $11.6 million and was producing an average of 864 b/d, mostly of a heavier grade of crude, at its Plaza Huincal oil field in the Neuquén territory. Given the importance of the gasoline market, the company decided that the Salta fields, which had tested a light crude, appeared to be the best prospect.[91] Its biggest competitor in Argentina at the time was the government company, Yacimientos Petrolíferos Fiscales (YPF), for which Congress already had reserved the most promising lands in Patagonia. YPF had established a refinery south of Buenos Aires, was selling petroleum products through nine hundred domestic outlets, and in 1926 was producing 13,000 b/d at its Patagonian fields, mostly of a heavier crude oil.[92] Despite YPF's production and the additional 11,700 b/d contributed by ten private companies, domestic consumption continued to exceed all state and private production of petroleum.

No such market existed in Bolivia, whose landlocked condition prohibited extensive export development unless the production of the Santa Cruz region could be combined with that of nearby Salta.

Jersey's prospectors early concluded that the export of Bolivian oil would require an 800-kilometer pipeline to the Paraguay River and thence a 1,100-kilometer barge ride to Buenos Aires.[93] Just the same, Jersey's geologists were enthusiastic about the oil possibilities in Bolivia and, based on the expected growth of the Argentine marketplace, Jersey Standard purchased two Bolivian concessions that consisted of more than 3.5 million acres from private parties. The Standard Oil Company of Bolivia established camps, built roads and facilities, and surveyed the entire holdings. Wildcatters discovered several deposits but, lacking a market, held back production to just 77 b/d by the end of the decade, sufficient only to support further exploration activities.[94] Operations were continued as an adjunct to Jersey's activities in Argentina.

In the process of searching for crude, Standard Oil of Argentina became a participant in the political struggle that was to make strong development by foreign enterprises impossible. Gen. Enrique Mosconi, head of YPF, singled out Standard as one of the foreign companies intent upon monopolizing Argentine petroleum resources. Basing his nationalist sentiments on an earlier refusal of Jersey's marketing agent in Buenos Aires to advance aviation gasoline to the Argentine military on credit terms, Mosconi accused the North American oil company of not working sincerely for Argentina's development but of controlling reserves for future exports.[95] The state government of Salta, however, welcomed Standard of Argentina as a company capable of filling the state treasury with royalty payments on properties that nineteenth-century mining codes had released to provincial control. The presence of the foreign oil company also promised lease payments to landowners, jobs for local workers, contracts for local businesses, advertising for local newspapers, and fees for lawyers. Salta was controlled by a group of nation-building politicians who, in this case, were intent upon boosting the economy of an interior province excluded from the previous wheat and beef growth that had benefited principally the litoral provinces.[96] Oil investment was the *salteños'* first experience with foreign capital, and the local oligarchy fought Mosconi and his friends in the national government at Buenos Aires to preserve its provincial prerogatives.

Mosconi and the other nationalists sought through a number of manuevers to restrict Jersey Standard's production activities. They succeeded first of all in securing larger tracts in the national territories for YPF's exclusive reserve. State monopoly of all subsoil wealth became an issue in the 1928 election, when ex-President Hipólito Yrigoyen returned to power in a campaign based on oil nationalism. His petroleum nationalization bill, however, languished in Congress for the next two years, a victim of political bickering, of salteño opposition, and of many a politician's reservations about injuring British interests along with North American. Having acquired some leases in Salta next to Standard of Argentina's oil fields, in the meantime, YPF began an offset campaign of its own. In 1929, Mosconi intervened to prevent the private company from securing a government permit to construct a pipeline that would carry crude produced by Standard of Bolivia into the Argentine marketplace. Finally, YPF attempted to undermine Jersey Standard's marketing position in the country by starting a price war that reduced gasoline prices and equalized them to consumers throughout the nation.[97]

There might have been an additional, technical factor, never mentioned in the nationalist rhetoric, that led Mosconi to neglect his Patagonian reserves while so assiduously moving YPF into Salta to break Jersey's Bolivian connection. YPF's volume production in the Patagonia was of a heavier crude that yielded little more than 5 percent gasoline before more extensive refining. Salta's and Bolivia's crude oil, on the other hand, was much lighter, naturally yielding a gasoline content of 40 percent and more, a product ideal for capturing and a market dominated now by the proliferation of automobiles and lorries across Argentina's roadways. Such an explanation accounts for Mosconi's zealous decision to carry on the struggle in Salta province.[98] Jersey Standard therefore, discovered before the 1920s were out, as it was learning in Mexico at the same moment, that its best markets in Latin America also were proving to be the most restrictive areas in which to expand production. While Jersey Standard maintained defensive positions in Argentina and Mexico and increased production only on existing properties in Colombia and Peru, its plans for expansion began to focus elsewhere in Latin America.

Shifting to Venezuela

Venezuela was an oil explorer's dream. Not having attracted much foreign investment in the late nineteenth century, the country had little of the economic infrastructure that made it a good market for petroleum products. The oil companies were to be Venezuela's first taste of fruits of progress. The dictatorship of Gen. Juan Vicente Gómez offered the foreign oilmen a welcome respite from the political obstacles that were hindering expansion in Peru, Mexico, Colombia, and Argentina. Curiously, there was no legal uncertainty over subsoil rights in Venezuela. Unlike the Mexican case, nineteenth-century liberalism had been too weak to have motivated repeal of colonial mining laws, so Venezuelans simply had not gotten around to "modernizing" their legal tradition. Also, much of the oil prospecting took place in frontier territories as yet sparsely settled by landowners. Thus, companies in Venezuela had understood clearly from the beginning that the petroleum deposits belonged to the state. Venezuelan government concessions of large size permitted the companies to avoid offset campaigns like those that had broken out on small Mexican parcels. Venezuela's proximity to the newly completed Panama Canal and to the Caribbean shipping lanes leading to refineries on the eastern seaboard of the United States enhanced its appeal to foreign oilmen. While roads, camps, and facilities had to be hacked out of the inhospitable tropics as in Colombia, Venezuela's oil resources presented fewer commercial transport problems certainly than did Bolivia's. The economic and political advantages of Venezuela were readily apparent to foreign oilmen.

Indeed, several oil companies already had established successful operations there by the time Jersey representatives first arrived in Caracas. Three Royal Dutch–Shell companies controlled the best land in the Maracaibo basin, and smaller British companies held coastal areas in the State of Falcón. These firms, by virtue of their early entry, operated under government concessions much more favorable than Jersey could hope to acquire in the 1920s.[99] Because the company initially chose not to purchase established producing properties, which were considered too expensive, Jersey failed to develop significant production in Venezuela until the end of the decade. Jersey men knew Venezuela had a favorable investment climate, the sort of environment that invited a company to be a petroleum pioneer.

Teagle sent out attorney T. R. Armstrong late in 1919 to apply for concessions from the Gómez government and to learn the Venezuelan system. At the time, the United States minister to Venezuela,

Preston McGoodwin, took seriously his instructions to aid American oil companies, passing word to Armstrong that the Gómez government might give preference to Jersey because of its "solid reputation and standing."[100] Armstrong must have been encouraged, for this was a sentiment rarely encountered in the hemisphere. He decided to obtain concessions according to the letter of the law, ignoring the advice of Jersey's sales manager in Caracas that getting concessions in Venezuela would take money and influence. An austere man, Armstrong especially objected to having to entertain lavishly and to paying intermediaries in order to gain favor.[101] It was to be by the book.

The American minister arranged for Armstrong to pay a courtesy call on General Gómez, a meeting that took place on the terrace of the president's private residence at Maracay. In a cordial mood, Gómez concluded by handing Armstrong a letter of recommendation addressed to the minister of development. Armstrong was convinced that his application for concessions would be approved at no additional expense other than those fees outlined in the petroleum laws.[102] Within a month, Armstrong formally submitted applications for five parcels south of Lake Maracaibo that Jersey geologists had been investigating. But these very same concessions went not to Jersey Standard but to Gómez' son-in-law, who promptly sold them to the Sun Oil Company![103]

Jersey men learned that success, after all, was achieved in Venezuela through intermediaries. The company eventually acquired about 1.2 million hectares of exploratory concessions. Some came from American companies dealing in Venezuelan oil concessions, while others were purchased through the brokerage of the president's son-in-law.[104] Jersey had two years and a year's extension in which to explore the widespread concessions. Several geological parties were sent out to appraise holdings in the Maracaibo basin, the mountains south of the lake, the Falcón coastal area, and eastern Venezuela. Using surface observations only in many difficult-to-survey areas, the geologists prepared maps for exploitation permits and for wildcat drilling.[105] By December 1921, the Standard Oil Company of Venezuela was incorporated in Venezuela, and Jersey now had the organizational presence to begin pumping and exporting crude, of which as yet there was none!

For years, Standard of Venezuela searched without success for a producing well. As a pioneering company, bringing in its own production, the subsidiary was proving a disappointment, for Jersey had spent nearly $14 million in Venezuela by the end of 1927 and had no production to show for it. At one place, a well was spudded in, and the oil promised to flow in commercial quantities. No sooner had a $250,000 holding tank been built nearby than the well went dry.[106] Much of the problem stemmed from the fact that Jersey men were retracing the paths of Shell geologists. In 1914, some two score Shell oilmen had begun scouting the Venezuelan hinterlands, securing for this competitor the most obvious prospects.[107] Despite its late start, Standard of Venezuela continued the wildcatting.

Unlike other places in Latin America, Standard's operations in Venezuela were spread throughout the entire country, and the dispersed camps and drilling sites compounded the work of the medical staff. The yearly malaria rate among workers in Venezuela was 741 per thousand, exceeding even that of Tropical in Colombia, which was 394 per thousand.[108] Wildcatting on the plains, in swamps, and in the mountains required the movement of heavy equipment by the most primitive means. To reach its destination at a remote drilling site, machinery had to be transshipped repeatedly between barges, mules, riverboats, and mules again.[109]

Despite the logistical and medical problems, the parent company supported Standard of Venezuela. The knowledge that other companies were pumping crude in increasing volume fortified corporate resolve. Venezuelan production rose from 18,249 metric tons in 1917 to 8,733,236 tons in 1927, with nearly all the production coming from the Maracaibo basin in western Venezuela and half of it from the subsidiaries of Royal Dutch–Shell.[110]

The favorable political atmosphere and the success of their competitors drove the Jersey men on, until in June 1928 wildcatters finally struck oil on a Standard concession. Working at a site in Quiriquire, eastern Venezuela, where no crude previously had been lifted, they sank rotary tools into a well that had been drilled unsuccessfully by cable four years earlier.[111] Through the application of the latest drilling technology, Jersey's efforts in Venezuela finally paid off by opening up a new zone for exploitation. Nowhere else in Latin America had the company felt politically secure enough to carry out such an expensive and exhaustive pioneering effort.

More important to its future producing position in Venezuela was Jersey's purchase, completed within days of the strike at Quiriquire, of the Creole Syndicate. The net value of the consolidated Venezuelan holdings came to $55 million, and the acquisition of Creole added a number of proven producing properties on the banks of Lake Maracaibo. Instantaneously, Jersey's share of the Venezuela's oil boom rose. The Creole connection brought with it a number of production contracts with concession holders near Shell's prolific oil fields in the west. The new Maracaibo production that began to increase soon after Jersey's drilling crews arrived had an additional advantage of being of a lighter-grade crude oil, yielding higher gasoline ends than Quiriquire's heavy asphaltic crude.[112] After a decade of work, Jersey Standard finally had established itself in Venezuela, not only as a pioneer oil company and but also as a buyer of producing properties.

Jersey and the Origins of Nationalism

How does one account for the success of the Standard Oil Company of New Jersey, despite its late start in Latin American production? Size obviously was important. For the most part, Jersey was able to finance the purchase and improvement of oil properties through the reinvestment of profits it made from its worldwide sales. Only once, in 1919, did it need to issue preferred stock to raise funds for expansion. The purchase of existing concessions allowed the company to enter quickly into oil development in Peru, Mexico, and Colombia, countries in which a formal request for exploratory concessions might have allowed nationalist politicians to delay its entry. In the cases of Argentina, Bolivia, and Venezuela, Jersey's subsidiaries were pioneers whose production growth was not as rapid, and indeed only in Venezuela was the production ultimately significant.

Jersey Standard also was a risk taker. With an eye to seizing production opportunities, Teagle and Sadler monitored the investment climate abroad as well as the activities of competitors. Some failures in Europe and in the Dutch East Indies had taught them the need for decisiveness, the result of which was a series of bold decisions that rapidly placed the company in control of production in Mexico, Peru, and Colombia. An additional strength was the company's competitive instinct. In view of the rise of the Shell group and the appearance of a number of competing marketers, Jersey executives no longer could be content with mere purchases of crude supplies from independent producers. Standard of New Jersey also had to have its own production abroad to hold its share of expanding petroleum markets. It was the success of Shell in Venezuela that drove Jersey Standard's explorers until they finally developed their own oil field.

Moreover, Jersey oilmen had the technological knowledge to exploit oil reserves in the most inaccessible regions. From their pool of twelve thousand employees worldwide, Jersey managers directed geologists, wildcatters, engineers, refiners, and pipeliners to the company's Latin American oil fields to eliminate production bottlenecks. Such experts applied their experience in North American oil fields to Latin American problems of exploration, survey, drilling, storage, transporting, training, health care, and administration.

Size, risk taking, and technological know-how were characteristics that other oil companies also possessed. Why Jersey Standard above the others? The explanation lies in the peculiar organization that this oil company inherited from the dissolution of 1911, for Standard Oil Company (New Jersey) above all had been a collection of refining and sales firms, and it retained its marketing organization that served the overseas and Latin American trade. Latin America came into Jersey's production purview because, as an oil company with few crude supplies under its direct control, it was vulnerable to competition from other foreign companies whose direct production in the hemisphere threatened Jersey's overseas markets. Standard's organizational structure also gave its producing affiliates access to worldwide markets that consumed nearly as much crude, heavy or light, as they could produce beyond the needs of the Latin Americans themselves. This is the reason that other independents sold their crude to Jersey affiliates.

In every way, Jersey Standard was a business success in Latin America's petroleum development, a success that drew attention. The company's reputation and importance made it an obvious target in the political battles that accompanied petroleum expansion. The industry had grown suddenly—quite before many governments had laws regulating it or even before politicians could agree on basic principles of ownership and taxation. The petroleum debate in Latin America began soon after the foreign companies started drawing crude oil from the ground.

Jersey Standard executives were unprepared to confront this opposition, and their actual political clout at any rate was limited.

Free debate of petroleum issues exposed men like Flanagan who attempted to keep secret their work for Jersey Standard and men like LeSueur who attempted heavy-handed resolutions of complex issues. Oil debates also embarrassed those nation-building politicians who had supported the company's efforts. Those countries with the longest history of foreign investment and of Jersey Standard imports provided the most inhospitable environments for Jersey expansion into production. Not even the United States State Department, for example, was inclined to support the letter of Jersey's contracts against strong nationalist opposition in Mexico.

Except in Venezuela, therefore, expansion outside the confines of the company's initial concessions was politically difficult. Most governments used their control of permits and concessions to increase their bargaining positions vis-à-vis the private companies. Mexican nationalists even had learned to split the united front of the oil companies, capitalizing on their natural competitiveness and gradually imposing more control over their activities. Quite often, these nationalistic measures inhibited growth of the petroleum industry in their countries, but considering that oil was a nonrenewable resource, few nationalists seemed unwilling to make the sacrifice. For all these reasons, even a company as economically powerful as Standard Oil Company (New Jersey) was politically vulnerable in Latin America.

The subsequent history of Jersey's subsidiaries stands as testimony to the gradual erosion of its position in Latin America. Encouraged by Argentina, the government of Bolivia expropriated Standard's operations there in 1937; the following year Mexico nationalized its foreign oil companies; and Brazil barred them from entering refining and exploration. Cuba took over the Jersey refinery and marketing apparatus in 1960. The operations of the International Petroleum Company continued to be a political issue in Peru until 1968, when the government expropriated it. Government relations in Argentina deteriorated to such a degree that the company's operations in Salta were pinched to extinction. Only in Colombia and Venezuela were conditions more favorable to Jersey. Yet the Colombian national petroleum company assumed control of the De Mares concession in 1951—exactly as the 1919 contract had stipulated. Jersey Standard was able to develop additional concessions in Venezuela as late as the 1950s, although this country also brought about nationalization, through the cooperation of the industry, in 1976.

Success of the nationalist movement, whose beginnings can be traced to the very arrival of the foreign companies, explains why Exxon today has more of a marketing than a production presence in Latin America. This may be changing once again. High world prices and financial shortages among the national companies, particularly in oil-importing nations, may lead to an expanded rule for private foreign capital in Latin American development. Exxon's participation in exploration, offshore drilling, and alternate energy plans seems to be pointing in this direction.

The subsequent national histories are influenced in large measure by growing state control over national resources and over the basic industries. In this process, each country has had its particular experience with foreign companies like Standard Oil Company (New Jersey). Each has its own oil history.

Notes

I would like to acknowledge a debt of gratitude to the late Henrietta M. Larson and to John D. Wirth and Alfred H. Saulniers for commenting on an earlier draft of this chapter.

The principal sources for this chapter derive from research materials collected in the late 1940s by Professor Larson. She was a member of the Harvard Business History team that researched and wrote the three-volume history of the Standard Oil Company (New Jersey). Some of these materials have appeared in the company histories cited in the notes, but much information on Latin America has never been published, and indeed many of the original company files have since been discarded. Although Professor Larson generously allowed me full use of her research, the interpretation remains my own.

Abbreviations

Creole	Records of Creole Petroleum Corporation, New York
Imperial	Imperial Oil Company, Ltd., Toronto
International	International Petroleum Company, Toronto
NA	U.S. National Archives, General Records of the Department of State
SONJ	Standard Oil Company (New Jersey), New York

1. It should be noted that foreign companies, for the most part, encountered more political opposition to their producing operations than to their

marketing and refining activities in South America. See Mira Wilkins, "Multinational Oil Companies in South America in the 1920s: Argentina, Bolivia, Brazil, Chile, Colombia, Ecuador, and Peru," *Business History Review* 48, no. 3 (Autumn 1974): 414–46.

2. George Sweet Gibb and Evelyn H. Knowlton, *The Resurgent Years, 1911–1927,* vol. 2 of *History of Standard Oil Company (New Jersey)* (New York, 1956): 455, 667.

3. See especially Adelberto J. Pinelo, *The Multinational Corporation as a Force in Latin American Politics: A Case Study of the International Petroleum Company in Peru* (New York, 1973), p. 3. My own research corroborates the conclusion of Neil H. Jacoby, *Multinational Oil: A Study in Industrial Dynamics* (New York, 1974), who states, "National sovereignty has prevailed over corporate power, notwithstanding the popular belief to the contrary" (p. 305).

4. The following information is based on the author's discussions with various executives at Esso Inter-America, Inc. (Coral Gables, Jan. 4, 1983); and on Exxon Corporation, *Annual Reports,* 1979–81; *Esso Inter-America, Inc.* (Coral Gables); *Intercambio,* n.s. 1, no. 3 (Oct. 19, 1982); *Boletín Informativo El Cerrejón, Zona Norte* 10, no. 1 (Dec. 1982); Exxon Corporation, *1981 Financial and Statistical Supplement to the Annual Report;* and *Petroleum Economist,* 1979–82.

5. Harold F. Williamson and Arnold R. Daum, *The American Petroleum Industry,* vol 1: *The Age of Illumination, 1859–1899* (Evanston, Ill., 1959): 643, 742.

6. Ralph W. Hidy and Muriel E. Hidy, *Pioneering in Big Business, 1882–1911,* vol. 1 of *History of Standard Oil Company (New Jersey)* (New York, 1955): 128, 259, 527; and Harold F. Williamson and Arnold R. Daum, *The American Petroleum Industry,* vol. 2: *The Age of Energy, 1899–1959* (Evanston, Ill., 1963): 251.

7. Hidy and Hidy, *Pioneering in Big Business,* pp. 486, 525.

8. A. H. Clarke, "History of the La Brea y Pariñas Estate—Peru," *Bulletin of the Pan-American Union* 28 (1909): 150; and Rory Miller, "Small Business in the Peruvian Oil Industry: Lobitos Oilfields Limited Before 1934," *Business History Review* 56, no. 3 (1982): 402. Miller's figures for Lobitos production are given in English tons, which the author has converted to barrels, one English ton equaling 8.75 barrels, according to material extrapolated from U.S. Senate, Foreign Relations Committee, *Investigation of Mexican Affairs, Report and Hearings,* 2 vols., 66th Cong., 1st sess., 1919, Senate Document 67, 1:270.

9. Senate Foreign Relations Committee, *Investigation of Mexican Affairs,* 1:209–10, 216–17; and Fritz Hoffmann, "Edward L. Doheny and the Beginnings of Petroleum Development in Mexico," *Mid-America* (Chicago) 24 (April 1942): 97–98.

10. Esperanza Durán de Seade, "Mexico's Relations with the Powers During the Great War" (D. Phil. thesis, St. Antony's College, Oxford University, 1980), p. 52; and George Philip, *Oil and Politics in Latin America: Nationalist Movements and State Oil Companies* (Cambridge and New York, 1982), p. 13.

11. Arturo Frondizi, *Petróleo y política*, 2nd ed. (Buenos Aires, 1955), p. 67.

12. Hidy and Hidy, *Pioneering in Big Business*, p. 713.

13. Bennett H. Wall and George S. Gibb, *Teagle of Jersey Standard* (New Orleans, 1974), pp. 72–74, 83.

14. Denny Ludwell, *We Fight for Oil* (New York, 1928), p. 33; and Gibb and Knowlton, *Resurgent Years*, p. 94. The Standard group in 1897 had attempted unsuccessfully to purchase the Royal Dutch Company. See Frederik Carel Gerretson, *History of the Royal Dutch*, 4 vols. (Leiden, 1953–57), 1:283–86.

15. Gibb and Knowlton, *Resurgent Years*, pp. 83, 335.

16. James E. Buchanan, "Politics and Petroleum Development in Argentina, 1916–1930" (Ph.D. diss., Dept. of History, University of Massachusetts, 1973), pp. 39–40; and Gibb and Knowlton, *Resurgent Years*, p. 182.

17. *Oil and Gas Journal* (May 2, 1919), pp. 54–55; and U.S. Department of State, *Foreign Relations of the United States, 1919* (1919), 1:167. Also see Gerald D. Nash, *United States Oil Policy, 1890–1964* (Pittsburgh, 1968), p. 49.

18. NA 23.6363/23, "Memo of the Office of the Foreign Trade Advisor," Apr. 8, 1920.

19. NA 821.6363/49, Philip Hoffman to Sec'y of State, desp. no. 8, June 5, 1919.

20. José Enrique Rodó's *Ariel*, 2nd ed. (Buenos Aires, 1966), first published in 1900, served for many years as the anthem of Latin American nationalists because it glorified Hispanic spiritualism, represented by Ariel, and denegrated the Anglo-American materialism of Caliban. Rodó was from Uruguay, a nation thoroughly transformed by foreign investment and scene of the first nationalist movement under José Battle y Ordóñez from 1903 to 1919.

21. Rosemary Thorp and Geoffrey Bertram, *Peru, 1890–1977: Growth and Policy in an Open Economy* (London and New York, 1978), pp. 50, 92.

22. U.S. Senate, *Investigation of Mexican Affairs*, 1:210; and Philip, *Oil and Politics in Latin America*, p. 28.

23. For some representative studies of this early nationalism, see Thomas F. McGann, *Argentina, the United States, and the Inter-American System, 1880–1914* (Cambridge, Mass., 1957); Robert Freeman Smith, *The United States and Revolutionary Nationalism in Mexico, 1916–1932* (Chicago and London, 1972); and Peter Seaborn Smith, *Oil and Politics in Modern Brazil* (Toronto, 1976), especially chapter I.

24. Irving Stone, "British Direct and Portfolio Investment in Latin America Before 1914," *Journal of Economic History* 37, no. 3 (Sept. 1977): 695. For North American and other foreign investments in Mexico, see Daniel Cosío Villegas et al., *Historia moderna de México*, vol. 7: *El porfiriato: La vida económica* (Mexico City, 1965): 1154.

25. *Foreign Relations of the United States, 1920* (1920), 2:213.

26. As quoted in Carl E. Solberg, *Oil and Nationalism in Argentina: A History* (Stanford, 1979), p. 19.

27. Gibb and Knowlton, *Resurgent Years*, p. 85.

28. Imperial, file 157, "Reports on London & Pacific Properties and Lobitos," April 22 and July 15, 1913.

29. International, reports of meeting of stockholders, 1913; and Secy's Office, Letters Patent, Sept. 10, 1914.

30. Imperial, Teagle Papers, files 20, 22, 36, 160.

31. Ibid., file 319, T. L. Scott to Milne & Co., Dec. 4, 1917; ibid., "Mr. Smith, Correspondence with Teagle," G. H. Smith to Teagle, Nov. 14 and 21, 1917; Wall and Gibb, *Teagle of Jersey Standard*, pp. 102–103.

32. International, Producing Dept., statistical tables; SONJ, Dept. of Economics and Statistics, statistical tables; Miller, "Small Business in the Peruvian Oil Industry," p. 411.

33. Imperial, Teagle Papers, files 10, 18, 87, 157, 160, 165, and 319.

34. Ibid., file 157, Milne & Co. to President of Peru, May 8, 1911; NA 323.415 in 8/1, Memorandum re La Brea y Pariñas, Le Sueur to U.S. Chargé, Lima, June 24, 1918.

35. Ibid.

36. Imperial, Teagle Papers, file 107-A, R. V. LeSueur to Milne & Co., Toronto, n.d.; SONJ, Legal Dept., Teagle Papers, "International Petroleum Company, Ltd.," Smith to Teagle, Nov. 14, 1917, and Teagle to LeSueur, New York, Nov. 27, 1917; and Wall and Gibb, *Teagle of Jersey Standard*, pp. 101–102.

37. Steve Stein, *Populism in Peru: The Emergence of the Masses and the Politics of Social Control* (Madison, Wis., 1980), p. 24; and Frederick B. Pike, *The Modern History of Peru* (New York, 1967).

38. NA 323.415 in 8/4, W. S. Penfield to Robert Lansing, Lima, Oct. 11, 1918. Canada and Great Britain, under wartime powers, commandeered the tankers of International and Lobitos.

39. SONJ, Legal Dept., Teagle Papers, "International Petroleum Co., Ltd.," correspondence, Sept. 1918.

40. NA 323.415 in 8/6, American consul-general to Dept. of State, desp. no. 663, Lima, Dec. 16, 1918.

41. NA 823.00/405, William E. Gonzales to secy of state, no. 695, Lima, Sept. 26, 1921; and NA 830.000/378, Gonzales to secy of state, no. 574, Lima, Jan. 12, 1921.

42. *West Coast Leader,* May 24, 1919, p. 1, as quoted in Gary R. Garrett, "The Oncenio of Augusto B. Leguía" (Ph.D. diss., Dept. of History, University of New Mexico, 1973), p. 117.

43. International, General Files, vol. 578, part 1 (1920–22).

44. Ibid., General Files, vols. 107, 533, 558, 578. Also see Peru, Ministerio de Relaciones Exteriores del Perú, *Tratados, convenciones y acuerdos vigentes entre el Perú y otros estados* (Lima, 1936), 1:386–93.

45. International, Statistical Dept., statistical tables; and ibid., General Files, vol. 305, part 2, "Peru. Labor conditions."

46. See George M. Ingram, *Expropriation of U.S. Property in South America: Nationalization of Oil and Copper Companies in Peru, Bolivia, and Chile* (New York, 1974), pp. 37, 60; and Pinelo, *The Multinational Corporation,* p. 155.

47. Esso Standard, Legal Dept., file 117, S. B. Hunt to M. F. Elliott, Dec. 3, 1913, C. O. Swain to E. Arredondo, May 7, 1915, and Arrendondo to Swain, May 8, 1915.

48. Antonio J. Bermúdez, *The Mexican National Petroleum Industry: A Case Study in Nationalization* (Stanford, 1963), pp. 2–3.

49. SONJ, Production Dept., Transcontinental Contract Files, Transcontinental Consolidated Oil Co., no. 7, Aug. 24, 1917.

50. SONJ, Directors' Files, Sadler Papers, E. J. Sadler to Hunt, Feb. 18, 1918.

51. U.S. Department of Commerce, Bureau of Foreign and Domestic Commerce, *Commerce Reports* (Sept. 13, 1920), p. 1223; Lorenzo Meyer, *Mexico and the United States in the Oil Controversy, 1917–1942,* trans. Muriel Vasconcellos (Austin, 1977), p. 4.

52. SONJ, minutes of meetings of Directors and Stockholders, 1919, 1927.

53. Ibid., Controller's Dept., reports of Transcontinental, 1918–28; ibid., Production Dept., Sadler's Mexican files, Sadler to Corwin, June 25, 1925.

54. Ibid., Coordination and Economics Dept., statistical tables.

55. Ibid., Controller's Dept., reports of Transcontinental.

56. N. Paulsen, *Map of Vera Cruz-Llave,* Aug. 1912; and *Carta de la zona petrolífera del norte de Vera Cruz,* March 1919.

57. Interview with L. Philo Maier, former Jersey engineer in Mexico in 1920, Coral Gables, Dec. 30, 1982.

58. SONJ, Sadler's Mexican files, "Mexico, 1920," J. A. Brown to A. F. Corwin, Oct. 27, 1920.

59. Ibid., Sadler's Mexican files, "Mexico, 1920," W. P. Haynes and Fred B. Ely to W. Warfield, Nov. 17, 21, 28, 1920; Jan. 6, 10, 1921; and May 1, 15, 1929.

60. Ibid., Sadler's Mexican files, "Mexico, 1920," Brown to Sadler, Apr. 29, 1920; Meyer, *Mexico and the United States,* pp. 50–51.

61. SONJ, Sadler's Mexican files, memorandum of Brown, Sept. 8, 1924, and telegram, Brown and Swain to Corwin, Sept. 18, 1924.

62. SONJ, Sadler's Mexican files, "Personnel, Finances, Legal," Sadler to P. E. Pierce, July 23, 1921, and Teagle to Sadler, Aug. 18, 1921.

63. Ibid., Sadler's Mexican files, "Personnel, Finances, Legal," Sadler to Teagle, Aug. 18, 1921.

64. On the Bucareli Agreement, see Wendell C. Gordon, *The Expropriation of Foreign-Owned Property in Mexico* (Washington, D.C., 1941), pp. 64–67; Meyer, *Mexico and the United States*, pp. 100–106; and John W. F. Dulles, *Yesterday in Mexico* (Austin, 1961), pp. 163–71.

65. SONJ, Production Dept., file 651, Lieb to Pierce, Apr. 1, 1927, and Lieb to Pratt, Dec. 28, 1927.

66. NA 812.6363/2289, "The petroleum controversy in Mexico, 1928," n.d.

67. SONJ, Production Dept., Sadler's Mexican files, "Personnel, Finances, Legal," Memo of C. O. Swain, April 4, 1927; ibid., Transcontinental files, "Free Lands, National Lands," R. Pratt to G. H. Lieb, July 20, 1927.

68. Samuel Flagg Bemis, *The Latin American Policy of the United States* (New York, 1943), pp. 217–18. Mexican historian Lorenzo Meyer views the Morrow-Calles agreements as "almost a complete victory for the Americans" in that the implementation of Article 27 further was delayed. See Meyer, *Mexico and the United States*, p. 135.

69. NA 812.6363/2558, Standard Oil Company (New Jersey) to Sec'y of State, April 27, 1928.

70. SONJ, Production Dept., "Transcontinental Land Dept.," R. Pratt, memorandum on Mr. Lieb's letter of June 21, 1928, about land policy in Mexico, July 9, 1928.

71. Ibid., Production Dept., Administrative files, "Early Development of the Oil Industry in Colombia," 1943; NA 821.6363/197, Minister H. Philip to secy of state, no. 783, Bogotá, Dec. 15, 1921.

72. SONJ, Production Dept., Corwin's Producing files, Teagle Papers, "Colombia . . . ," reports on Colombia, 1919.

73. NA 821.6363/131, Bogotá Legation to secy of state, Jan. 29, 1921; SONJ, Production Dept., "Administration, Legal, Colombia," T. R. Armstrong to Corwin, April 20, 1920.

74. Ibid., Production Dept., file 8.280, "Colombia Concessions—Demares," translation of contract of Aug. 24, 1919; and Colombia, Ministerio de Obras Públicas y Fomento, *Memoria del Ministerio de Obras Públicas al Congreso de 1919* (Bogotá, 1919), pp. 13–15.

75. International, vol. 818, "Colombia. General letters of progress . . . "; SONJ, Production Dept., "Corporate, Colombia," 1920–21; A. M. McQueen, "Outline and Developments on Colombia Concession of Tropical Oil Company," as cited in Colombia, Bureau of Information, *Colombia Review* (Nov. 1927), pp. 204–205.

76. SONJ, Production Dept., "Colombia," Condition of Operations, June 30, 1921; ibid., Production Dept., "Tropical Oil Company Development during 1926."

77. International, file 818, "General letters of progress," 1922 and 1927; and NA 821.6363/215, Consul E. C. Soule to secy of state, Cartegena, March 16, 1922.

78. International, file 870, "Labour Matters General," H. A. Metzger to G. H. Smith, June 16, 1926.

79. Ibid., file 817, "Medical Matters General"; A. W. Schoenleber to G. H. Smith, Sept. 2, 1921; ibid., "Tropical Oil Company, Annual Report," 1926.

80. SONJ, Controller's Dept., Consolidated Div., report of Tropical Oil Co., 1927.

81. International, file 878, "Geological Surveys—Seismic Work," S. B. Hopkins to A. M. McQueen, Jan. 20, 1926; Charles E. Kern, "Oil Possibilities of Colombia Are Interesting Many Companies," *The Oil and Gas Journal* (Oct. 14, 1926).

82. Stephen J. Randall, *The Diplomacy of Modernization: Colombian-American Relations, 1920–1940* (Toronto, 1977), pp. 12, 57; *El Espectador*, (Bogotá), June 5, 1927, p. 3.

83. SONJ, Teagle Papers, "Colombia, Tropical Oil Co., Andean Nat'l Corp., Ltd.," memorandum referring to oil pipeline by Flanagan, Sept. 1919.

84. Imperial, file 807, Correspondence with Flanagan, letter from Carlton Jackson, Feb. 22, 1922.

85. Ibid., File 807, Correspondence with Flanagan, Flanagan to Brock, Apr. 27, 1920, and various letters of Flanagan, Aug. 1920 and May 1921.

86. The controversy is recorded in letters and clippings from various Colombian newspapers in Ibid., file 807, Correspondence with Flanagan; and SONJ, Teagle Papers, "Colombia."

87. Ibid., Production Dept., file 8.300, "Legal," address delivered by Flanagan, Bogotá, April 27, 1928; Colombia, Ministerio de Industria y Trabajo, *Anexos a la "Memoria de 1925"* (Bogotá, 1925), pp. 90–93; "Petroleum Transport in the Tropics," *The Imperial Oil Review* 10, no. 5 (Sept. 1926), pp. 2–10.

88. Randall, *The Diplomacy of Modernization*, pp. 101–105; NA 821.6363/781, U.S. Embassy to State Dept., no. 800, Dec. 28, 1929; NA 821.6363/679, no. 485, Embassy to State Dept., no. 485, Aug. 12, 1929.

89. SONJ, Production Dept., Corwin files, report of T. R. Armstrong on the Argentine, April 19, 1926.

90. Ibid., Production Dept., Corwin files, memorandum of Sadler on the Argentine, May 25, 1923, and press release of Standard Oil of Argentina, Aug. 19, 1927.

91. Ibid., Economics and Coordination Dept., statistics of Standard Oil of Argentina production.

92. Ibid., Production Dept., Corwin's files, report of Armstrong, Apr. 19, 1926. For the development of YPF as a competitor of the private oil companies, see Solberg, *Oil and Nationalism in Argentina*; and [Roy Leigh],

"Foreign Producers in Argentina Facing Uncertain Government Policy," *Oil Weekly* 55 (Dec. 13, 1929): 73.

93. SONJ, Production Dept., Corwin's files, "Bolivia," Corwin to Teagle, Sept. 23, 1918.

94. Ibid., Production Dept., Corwin's files, "Bolivia Braden Concessions, 1920/21," memos by Corwin and Warfield, Nov. 19, 1920; Ibid., Economics and Coordination Dept., statistics on Bolivia.

95. Enrique Mosconi, *El petróleo argentino, 1922–1930, y la ruptura de los trusts petrolíferos inglés y norteamericanos . . .* (Buenos Aires, 1936), pp. 16–18; Solberg, *Oil and Nationalism*, pp. 104–105.

96. Frederick Alexander Hollander, "Oligarchy and the Politics of Petroleum in Argentina: The Case of the Salta Oligarchy and Standard Oil, 1918–1933" (Ph.D. diss., Dept. of History, University of California, Los Angeles, 1976), pp. 302–463.

97. Ibid., p. 291; Frondizi, *Petróleo y política*, pp. 239, 242; and especially Carl Solberg, "Entrepreneurship in Public Enterprise: General Enrique Mosconi and the Argentine Petroleum Industry," *Business History Review* 56, no. 3 (Autumn 1982): 380–99.

98. Crude at Comodoro Rivadavia had an American Petroleum Institute rating of 20° while Standard of Bolivia's new wells at Bermejo tested at 30° to 60° A.P.I. See [Roy Leigh], "Prospecting in Southern Bolivia and Northern Argentina," *Oil Weekly* 55 (Dec. 13, 1929): 142; "Analyses of Typical South American Crudes," *Oil Weekly* 55 (Dec. 13, 1929): 196–208. Carl Solberg first questioned Mosconi's oil strategy in his "Entrepreneurship in Public Enterprise," pp. 390–91.

99. Edwin Lieuwen, *Petroleum in Venezuela, a History* (Berkeley, 1955), p. 15; Rómulo Betancourt, *Venezuela: Oil and Politics*, trans. Everett Bauman (Boston, 1979), pp. 14–16.

100. Creole, Armstrong file, J. H. Senior to Corwin, June 17, 1919.

101. Ibid., Armstrong file, Armstrong to Corwin, Dec. 30, 1919.

102. Ibid., Armstrong file, Armstrong to Corwin, Dec. 13, 1919.

103. Ibid., Armstrong file, Armstrong's notes on conference with Julio J. Méndez, May 18, 1919, and with Preston McGoodwin, May 19, 1919, and Armstrong to Corwin, May 24, 1920.

104. Ibid., "Concessions—Sanabria" and "Concessions—Méndez (Kunhardt)," memorandum, Sept. 17, 1920.

105. Ibid., progress report, H. F. Dawson to Sadler, Nov. 12, 1920; ibid., reports on Venezuela (general), Armstrong to Sadler, Feb. 21, 1921, F. A. Dalburg to Armstrong, Feb. 7, 1921, and reports of survey, Oct. 12, 1911.

106. Ibid., Reports on Venezuela, Rutherford Bingham to William Warfield, Sept. 28, 1922; and ibid., "Concessions—Los Barossos," Dalburg to Dawson, Oct. 5, 1921.

107. See Ralph Arnold, George A. Macready, and Thomas W. Barrington, *The First Big Oil Hunt: Venezuela, 1911–1916* (New York, 1960).

108. Creole, Venezuela medical reports (1924).

109. Ibid., reports on Venezuela, inspection trip to Southern Paija, May 19, 1924.

110. U.S. Tariff Commission, *Report to the Congress on the Cost of Crude Petroleum*, 2nd series, no. 4 (1931): 19; and Lieuwen, *Petroleum in Venezuela*, p. 40.

111. Anibal R. Martínez, *Chronology of Venezuelan Oil* (London, 1969), p. 57.

112. Henrietta M. Larson, Evelyn H. Knowlton, Charles S. Popple, *New Horizons, 1927–1950;* vol. 3 of *History of Standard Oil Company (New Jersey)* (New York, 1971), p. 42; [Roy Leigh], "Venezuela Has Ten Potential Oil Fields," *Oil Weekly* 55 (Dec. 13, 1929): 77–78.

Carl E. Solberg

YPF: The Formative Years of Latin America's Pioneer State Oil Company, 1922–39

Argentina assumes major importance in an analysis of Latin American petroleum history for two reasons. First, Argentina is one of the leading oil-producing countries in the Latin American region. Indeed, it ranks third—behind Venezuela and Mexico—in oil output. Although it is not yet an exporter, by 1980 Argentine production covered the nation's large oil demands almost completely, and the potential for increased production in the future is good. While Argentina is a large producer, it also is an old one, and the age of the oil industry points to the second reason why the Argentine experience is significant. Soon after oil was discovered (in 1907), the government entered the oil business, and in 1922 the government created the first state-owned oil company in Latin America: Yacimientos Petrolíferos Fiscales (YPF).[1] Because of YPF's pioneering role—and because several other countries initially modeled their state oil companies along the lines that YPF had adopted—the Argentine company's political and economic problems presaged much of what was to occur elsewhere in the Latin American petroleum industry. Today, however, YPF is atypical of the Latin American state-owned oil company. Unlike state companies in exporting countries like Venezuela and Mexico or in Brazil, which is an oil importer, YPF does not enjoy monopoly control over major sectors of the oil industry, including production, refining, and marketing.[2] And, what is more, YPF is one of the financially least successful of the Latin American state companies. While Argentina has large oil resources, YPF remains unable to develop their full potential.

While Mexico and Venezuela—and to a certain extent also Brazil—have developed an oil-policy consensus, Argentina has not.

SURTIDOR SIAM-BLOK "88"

El preferido por el público automovilista
y el más hermoso de los surtidores de nafta

SIAM DI TELLA L^{TDA}
AVENIDA DE MAYO 1302

U. T. 35, LIBERTAD 4041   BUENOS AIRES

This advertisement by S.I.A.M., producer of YPF's gasoline pumps,
indicates the close link between Argentina's state oil company
and the nation's nascent manufacturing sector. From *Yacimientos
petrolíferos fiscales*, 1934.

For over seventy years now, Argentines have been deeply, often bitterly, divided over the role the government should play in the oil business. Some would like to dismantle YPF and turn the oil industry entirely over to private enterprise. Others would like to exclude all private investment, especially foreign investment, from the oil industry and grant YPF a monopoly. The basic issue is this: What is more important, rapid production growth or national ownership of the industry?

While this dispute rages, the state company hobbles along from one crisis to another. And this sorry state of affairs will, in all likelihood, continue until Argentina's political crisis is resolved on a more fundamental level, for this country, so rich in resources and so full of promise, is a classic conflict society in which agreement on basic political and economic issues is absent. When—or if—Argentina will solve its political problems is impossible to foretell, but until Argentines agree on fundamental policy directions, the future of YPF and of private oil investment in Argentina remains an open question.

This essay examines the experience of YPF during its formative years, from 1922 to 1939. In this period the company emerged from obscurity to become the largest oil producer in Argentina; but it did not become a monopoly, as did the state oil company in Mexico at the same time. In this analysis, I will focus on why YPF was created, its goals and objectives, the policy disputes that surrounded it, and the terms of the petroleum legislation that shaped its future. My fundamental objective will be to discover why the Argentine state company, which had an early start and enjoyed rapid growth at first, remained financially weak and continued to share production, refining, and marketing functions with the private oil companies.

Argentina's Oil Industry Today: An Overview

A closer look at the Argentine oil industry in recent years will give perspective for analyzing the issues that surrounded YPF's formative period. At the end of 1978, Argentina's proven oil reserves stood at 380 million cubic meters (or 2.4 billion barrels), which made them the third-largest reserves in Latin America although much smaller than those of Venezuela and Mexico.[3] These petroleum resources lie in roughly a north-south belt extending the length of the country along the eastern slopes of the Andes from the Bolivian border in the north to Tierra del Fuego in the far south; in Patagonia, where the country is much narrower than in the north, the main oil fields are near the Atlantic coast. The main producing areas along this belt are in Salta and Jujuy provinces (adjacent to Bolivia), in Mendoza and Neuquén provinces (in central Argentina near the Chilean border), and in the Patagonian provinces of Chubut and Santa Cruz, where the country's first oil field (and long its most productive) is centered around the city of Comodoro Rivadavia. Although potential exists for increasing production in these areas, particularly in Neuquén and Salta, in recent years the principal focus of exploration has been in the province of Tierra del Fuego— especially in the offshore areas of this southern tip of Argentina. In the mid-1970s, when American geologist Bernardo F. Grossling raised the possibility that "giant-size accumulations like the Middle East" might exist on the Argentine continental shelf and might prove four times the size of the oil deposits on the United States continental shelf, oilmen became very interested in the Patagonian offshore region.[4] Such estimates led the *Wall Street Journal* in 1981 to estimate that Argentina's potential reserves were "at least 10 billion barrels" and fostered some heady thinking in Argentina. One recent essayist speculates that Argentina may become the world's largest oil-producing country, and the military regime that seized power in 1976 formulated its oil policy with the aim of making Argentina a major petroleum exporter.[5]

But there are very serious problems and obstacles to confront before such hopes can be realized. For one thing, in the light of more recent evidence, the reports of Grossling and others may be unduly optimistic, especially in the context of the enormous investments necessary to explore and produce oil in an area of notoriously tempestuous weather. To obtain a thorough evaluation of the oil resources of the Malvinas basin, Argentina needs the cooperation of the international oil companies, but here the problem of offshore sovereignty intrudes.[6] Argentina has one difficult, and potentially dangerous, dispute with Chile over the sovereignty of three islands in the Beagle Channel off southern Tierra del Fuego. But this problem recently has been overshadowed by the claims of Britain, which emphasizes that Argentine exploration off the Strait of Magellan violates its claims to sovereignty in waters surrounding the Falkland Islands (or the Malvinas, as they are known in Latin America).

Whether or not possible oil deposits played a major role in the recent Falklands War, Britain and Argentina had been engaging in a diplomatic sparring match over offshore sovereignty since 1976. In 1981, Argentina dismissed British protests against offshore exploration as "irrelevant," while the British ambassador to Argentina asserted that negotiations over the Malvinas' sovereignty were "practically at the bottom of the list" because of the region's presumed oil and fishing resources.[7] The bottom line is that Argentina's defeat has dealt its more optimistic offshore oil hopes a major setback, and until a long-term political settlement with Britain over the area is reached, Argentina will have to concentrate its petroleum exploration and production efforts onshore and in offshore areas outside British claims.

Another serious political stumbling block—this one domestic—will delay the intensification of exploration and production in Argentina. Defeat in the Falklands War also reopened Argentina's long-standing internal political crisis, signaling a major shift in Argentine oil policy. The officer corps had played a strong role in the formative years of YPF, but the military became increasingly dissatisfied with the results of nationalist petroleum policies by the 1960s. Thus, the military governments that ruled Argentina for most of the period after 1965 favored private enterprise in the oil sector. This policy orientation became particularly pronounced during the regime of Gen. Jorge Videla, in power from 1976 to 1981. Videla's government was ideologically committed to free-market capitalism and reduced the role of the state-owned companies (although not the military's own manufacturing establishments).

In the petroleum sector the Videla government whittled away at YPF's position and placed men in control of YPF who were openly hostile to the company's historic paramount position in the industry. Meanwhile, the government welcomed foreign oil investment, and it also encouraged private Argentine firms to enter petroleum production. Previously, petroleum capital had been provided either by the state or by foreign oil companies, but by the 1970s, ambitious Argentine entrepreneurs were becoming eager to share in the nation's beckoning oil bonanza. Domestic Argentine contractors, many of them engaged in secondary recovery operations, nearly tripled their output between 1975 and 1978, when it reached 73,000 barrels per day (b/d).[8]

Meanwhile, YPF's financial position became critical. It lost money every year after 1975, and by the end of 1981 the company had accumulated a $4.1 billion debt.[9] YPF's losses resulted, at least in large part, from policy decisions beyond its control. Since 1976, the government kept oil price increases well below the inflation rate, which meant that YPF's costs rose faster than its income.[10] The state company also had to bear an extremely heavy tax burden. By the mid 1970s, YPF paid 68.4 percent of its sales revenues as taxes to the national government, provincial governments, and a long list of state agencies, including the National Energy Fund, the National Highway Council, and the Provincial Road Fund. Another serious financial drain resulted from the failure of other government enterprises, such as the National Railways and the state airline Aerolíneas Argentinas, to pay YPF for their fuel. These intragovernmental accounts represented 20 percent of YPF's total sales and were as much as ten years in arrears! As a result of its weak financial position YPF lost many of its best experts, due to low pay, and its exploration funds have been inadequate, causing production growth to lag.[11] As we shall see, in its tax policy the military regime was following a path that previous civilian governments had taken—a path that regarded YPF as a convenient source of revenue to ease the Argentine state's serious and chronic deficit position.

The military regime justified its antistatist oil policy by pointing to the steady increase in Argentine oil production after foreign investment had been allowed to enter the oil industry in 1976.[12] And indeed, production had increased from 23 million cubic meters per day in 1975 to 29 million in 1981 (at this point Argentina was about 92 percent self-sufficient). And the military regime was also concerned over estimates that Argentina would have to invest huge sums—the *Petroleum Economist* put them at $60 billion between 1982 and 2000—to avoid a big drop in oil production late in the century.[13]

In December 1981, a new military government headed by Gen. Leopoldo Galtieri took office and immediately began to make plans to increase foreign oil investment and to downgrade YPF's role still more. President Galtieri, an admirer of United States President Ronald Reagan, was also much enamored with the free-enterprise, deregulated model of capitalism currently being preached in the United States by the Republican administration, and the Galtieri administration had high hopes for the benefits that Argentina would receive as a result of formulating its foreign policy as well as its domestic economic policy in tandem with Reagan's America. In terms of oil, Galtieri planned to deregulate the Argentine petroleum industry. Late in 1981 and early in 1982 rumors abounded in Buenos Aires that the government was planning to dismantle YPF and turn it over to the private sector. This turned out not to be quite the case, for the government's draft Hydrocarbons Law, whose terms became known in March 1982, did not abolish YPF but instead condemned it to a slow and lingering death. The government planned to limit YPF to its present producing fields (and possibly to turn some of them over to private enterprise), to open all new exploration and production to private capital (domestic or foreign), to allow the private companies to compete freely for the domestic market, and to permit crude prices to rise to international levels. Under these circumstances, Argentine oil production probably would have risen, but YPF would have lost its primacy in the Argentine oil industry, and the state company might have been doomed.[14]

But the political uproar that shook Argentina following the Galtieri regime's disastrous defeat in the Falklands War suddenly put this petroleum policy on ice. Military rule in Argentina had suffered a decisive, perhaps fatal setback. The generals, who had presided over a rapidly deteriorating economic and social situation, had lost whatever little public legitimacy they had left, and Galtieri's government fell only six months after assuming power. The new administration of Gen. Reynoldo Bignone promised to hold elections and return Argentina to civilian rule, and by December 1983 this was done. Today, it is difficult to see how any military government, unless it took a markedly nationalistic and populist stance, could gain mass political legitimacy in contemporary Argentina.

Bignone dropped Galtieri's proposed oil law, and YPF has survived intact. "Let us not fool ourselves," lamented the *Review of the River Plate*, a staunch supporter of private oil capital. Under the present political circumstances, "clearly the proposed sub-soil rights law can be forgotten for a long time."[15]

Thus, ironically, the Malvinas have given YPF a new lease on life. And it is reasonable to assume that the oil policy of the new civilian government of Raúl Alfonsín and his middle-class Unión Cívica Radical will favor YPF more and the private petroleum sector less than Argentina's recent military rulers. I will now turn to the early history of YPF to examine the formative years of the state company that became a symbol of Argentine economic nationalism and managed to survive decades of political adversity.

The Origins of the State Oil Industry, 1907–30

Argentina at the turn of the century produced no oil, but the nation was enjoying an agricultural export boom that already had made it the leading economic power in Latin America. Since colonial times Argentines knew that petroleum existed in Salta and Mendoza provinces, but the nineteenth-century attempts that a few intrepid Argentine capitalists made to produce oil failed. This situation did not unduly alarm the proud and wealthy landed oligarchy that long had controlled the Argentine government, for the nation had become something of an economic satellite of Britain, financed by British capital and tied to British (and other European) export markets. The British connection traditionally had assured Argentina a cheap source of fuel in the form of Welsh coal that entered Argentina duty free and enjoyed favorable freight rates on the British ships that carried most of Argentina's foreign trade. Argentina also imported substantial amounts of petroleum through subsidiaries of the two multinational giants, Royal Dutch–Shell and Standard Oil of New Jersey. This imported coal and oil powered the nation's extensive railway system as well as its electricity plants, meat-packing factories, and military and naval installations.[16] Because Argentina produced no coal and because its petroleum resources were almost entirely unknown, foreigners as well as most Argentines believed that South America's second-largest nation had little industrial future beyond the processing of its agrarian production.

This all began to change in 1907, when government drillers struck oil on state-owned land near the Patagonian town of Comodoro Rivadavia. Whether the drillers were searching for water, as they claimed, or whether the government was secretly looking for oil has been a matter of dispute ever since, but in any case the next day President José Figueroa Alcorta proclaimed the area surrounding the discovery a state oil reserve. He was able to take this decisive action because Comodoro Rivadavia (and indeed all Patagonia at this time) was located in the national territories. By the 1853 Constitution and subsequent legislation, the central government possessed the authority to acquire and administer mineral

concessions in the territories. As we shall see, the federal government did not possess this authority in the provinces, a situation
that would lead to conflict between the provincial administrations
and the central government as the oil industry developed.

Soon after the 1907 discovery, the federal government began to
explore its Comodoro Rivadavia reserve and to produce a little
crude, which the local state-owned railways utilized. This entrance
of the government into the oil business touched off a political debate that soon became bitter and that has surrounded the Argentine
oil industry ever since. Powerful political leaders representing the
agro-exporting landed elite argued that free-market capitalism was
the key to Argentina's prosperity. Private enterprise and foreign investment, argued these politicians, were the sources of Argentina's
progress, and in the oil industry, the private sector would bring
equally spectacular results. The government should stay out. These
leaders fought every step of the way attempts by President Figueroa
Alcorta, who held office until 1910, and by his successor, Roque
Sáenz Peña, to appropriate funds for the government operations. As
a result of this opposition, the government reserve at Comodoro
Rivadavia was reduced greatly (to 7,950 hectares), and appropriations to support the state's drillers were sufficient to produce only a
trickle of oil. In fact, the total federal budget support that the state
oil industry received—and it all came between 1910 and 1916—
was only 8.6 million paper pesos, or about $3.6 million U.S. dol-
lars. That the government's operations survived at all was largely
due to President Sáenz Peña. Although he was in general an advocate of laissez-faire capitalism, he was also something of a yankee-
phobe, and he feared that if the government abandoned the oil business, the American oil companies would soon come to dominate it.
This fear, along with the disruptive impact that British miners'
strikes made on Argentine coal supplies, led Sáenz Peña to establish Latin America's first state oil company in 1910. This predecessor of YPF was originally a dependency of the Ministry of Agriculture.[17]

Convinced that Argentine politics had to represent the newly
emerging middle classes, Sáenz Peña also reformed the nation's
electoral laws, and as a result, in 1916, the first democratic presidential election in Argentine history brought Hipólito Yrigoyen,
leader of the middle-class Unión Cívica Radical (Radical Party), to
power. Yrigoyen, who carefully and successfully cultivated an im-

age of champion of the common people, was the central figure in Argentine politics between 1916 and 1930.

But this political change did not initially signal a policy change for the struggling state oil company. During his first term (1916–22), Yrigoyen generally followed traditional liberal economic policy. He did little to promote industrialization. Although the Radical Party breathed nationalist rhetoric, it was in fact closely tied to pro-British agrarian exporting interests, and, rather than support government oil development, Yrigoyen chose to wait until the end of World War I would again provision Argentina with cheap fuel. Despite the fact that Yrigoyen ended his predecessors' policies of supporting state oil development with budgetary appropriations, high wartime fuel prices did bring profits to the government oil fields, enabling them to expand production considerably. Nevertheless, during the first Yrigoyen presidency, the operations at Comodoro Rivadavia were plagued with equipment shortages, labor unrest, and manifold bureaucratic and political interference.[18]

Yrigoyen's economic policies evoked protest as the economic significance of the war became clear. By the early 1920s, Alejandro E. Bunge, a prominent economist, began to challenge the viability of Argentina's export-oriented economy and to urge industrialization and more self-sufficiency.[19] Arguments like this found a receptive audience among an influential group of army officers who deplored the severe fuel shortages Argentina had suffered during World War I. The war had shown, they argued, that during times of crisis Argentina could not count on adequate British coal supplies. In the name of national defense, they urged the government to support the creation of an industrial structure and, above all, to make the nation self-sufficient in petroleum. Although some officers did not fear foreign oil capital, the most vocal and influential of the nationalists insisted that the state must own the oil industry. They also envisaged a role for the army in leading the new state enterprise.[20] Thus, when the military first took a role in Argentine oil policy, it was a nationalist position. Nonetheless, these army officers of the 1920s were not opposed to private capitalism in general. Like public entrepreneurs in other Latin American countries, notably Brazil and Mexico, the Argentine "entrepreneur generals," as José Luis de Imaz has called them, aimed to create an infrastructure of state industry in petroleum, steel, and other basic sectors that would promote the growth of privately owned manufacturing.[21]

The petroleum nationalists in the army were not isolated, for civilian opinion was also becoming critical of foreign oil investment. The American oil companies, Jersey Standard in particular, had suffered from a poor public image in Argentina during the World War I period. Just before the war, Luis A. Huergo, a distinguished engineer whom Sáenz Peña had named to direct the government oil fields, issued a famous manifesto attacking Standard Oil as an "octopus" and a "band of cruel, usurious pirates." During the war (and during the Mexican Revolution, then in progress), numerous influential journalists and politicians charged that Standard was interfering in Mexican politics and argued that Argentina would suffer the same fate if Standard were to expand its operations there. As Ricardo Oneto, a widely read journalist and champion of the state oil industry put it in 1916, "Mexico, a victim of the wealth of its oil resources and of the voracity of the trusts, ought to serve as a lesson to the statesmen of countries like Argentina that still enjoy control over their oil resources."[22] In 1917, a group of congressmen accused Standard of tariff evasion, and in 1919 a congressional antitrust committee, which generated great public interest, severely criticized the marketing policies of Standard's importing subsidiary as monopolistic.[23]

Nationalists were becoming particularly concerned because Jersey Standard, which previously had imported oil rather than producing it, was showing signs of interest in Argentine oil production by the end of World War I. The company acquired petroleum concessions in Neuquén territory and in Salta province, and in 1922 Jersey Standard incorporated a wholly owned Argentine production subsidiary. Standard was the only important American company seriously interested in Argentine production, but it was by no means the only private firm that began drilling. The two British giants, Shell and Anglo-Persian, both acquired concessions in Patagonia soon after the 1907 discovery. Anglo-Persian's holdings proved unfruitful, but Shell by the 1930s became a significant producer. There were private Argentine oil companies too. One of them, the Compañía Argentina de Comodoro Rivadavia, possessed a rich Patagonian concession, but in 1920 it leased its holdings to a company belonging to the British-owned Argentine Railways. The private firm of Astra, which began to produce in Patagonia in 1916, was a successful company, but it too passed out of Argentine control and into the hands of German investors in the 1920s.[24] Although several small private Argentine companies did exist, Argentine capitalists avoided the high-risk oil business and, in a pattern that had been well established for decades, concentrated their investments in pampa landholdings, which were lucrative, secure, and prestigious.

When Yrigoyen renamed the state oil company Yacimientos Petrolíferos Fiscales just before he left office in 1922, influential sectors of military as well as of civilian opinion were becoming deeply concerned about the investment pattern in the oil industry. Private Argentine holdings were marginal, and British concessions were confined to Patagonia. But Jersey Standard was rapidly on the move. Many Argentines feared that, if left unchecked, the giant American corporation would soon dominate the industry. The administration that took office in 1922 would mobilize this nascent petroleum nationalism behind YPF.

Enrique Mosconi in Charge of YPF

Yrigoyen could not constitutionally succeed himself, although he could run for the presidency again after an interval of one six-year term. His successor, Marcelo T. de Alvear, was also a Radical and was elected with Yrigoyen's support. But Alvear soon made it clear that he intended to run the government his own way and not necessarily as Yrigoyen wished. Alvear's de facto declaration of independence not only sharply divided the Radical Party; it also signaled new policy directions in several areas, including petroleum. While Yrigoyen's oil policy had been ambivalent and had given only lukewarm support to the state oil enterprise, Alvear made development of the newly named YPF one of his administration's top priorities. And to head the state oil company, Alvear appointed a most remarkable military man, Col. Enrique Mosconi.

Mosconi was the most important individual in YPF's history. Born in 1877, this son of an Italian immigrant had pursued a successful military career, including advanced training with the German Army. While subdirector of the Argentine Army War Arsenals between 1914 and 1919, Mosconi became well acquainted with the problems of provisioning the armed forces in a country that had little industry. Convinced that national defense required an industrialized economy, he became one of the most outspoken economic nationalists in the army officer corps. After the war, Mosconi served as director of the Army Aeronautic Service, and in this post he was a pioneer in Argentine military and civilian aviation development. He also became acutely aware that military aviation in Argentina depended on fuel that could only be obtained from foreign suppliers.[25]

The ambitious and energetic Mosconi was described at the time as a "live wire" who had "no time for ornamental people who stand around and do nothing."[26] And as director-general of YPF between 1922 and 1930, Colonel (General after 1926) Mosconi certainly lived up to that reputation. He effectively reorganized the demoralized and confused state oil company, set ambitious goals for YPF, and moved a long way towards carrying out his plans. What is more, Mosconi was a skillful public relations man who created an effective image of YPF as friend and defender of the Argentine consumer. More than anyone else, Mosconi was responsible for the identification of YPF with Argentine dignity and pride. Mosconi also had his share of faults and weaknesses. He was a stubborn and somewhat self-righteous man who could be arrogant towards his scientists and impatient with politicians who got in his way. In these respects, he was a typical pioneer oilman, and in fact some of his admirers have compared him to John D. Rockefeller or Sir Henri Deterding.[27]

His staunch opposition to the foreign oil companies made the man and his memory an embarrassment to conservative governments of the 1930s and later, but Mosconi nonetheless has become something of a popular hero in Argentina. And Argentine nationalist intellectuals have created a virtual Mosconi hagiography. Ricardo Oneto saw Mosconi as "the great builder with the statesman's penetrating view of the future." More recently, Jorge Scalabrini Ortiz celebrated Mosconi as "that great patriot who saved Argentine oil for the nation." One recent writer went so far as to say that the "petroleum general" "was the only figure who knew how to interpret our machismo (sentimiento machista) when he declared war on the trusts."[28]

The goals and objectives that Mosconi established for YPF deserve comment, for they were more complex than these paeans of nationalist praise indicate. Like many military men of his generation, Mosconi believed that industrialization and self-sufficiency in basic industrial products were essential in the name of national defense. And since petroleum was the basic for all modern industrial development, a thriving oil industry was critical to Argentina's economic progress and independence. The foreign oil companies could not be trusted to respond to Argentina's needs; the national government, therefore, would have to establish a YPF monopoly over the oil industry. Mosconi, however, thought it would be unwise and impractical to expropriate the existing properties of the foreign oil companies. Instead, they should be limited to the holdings they possessed.[29] The state oil firm should become a vertically-integrated enterprise that would produce oil, refine it, and market a wide range of petroleum products at prices low enough to stimulate industrial growth and to make the average consumer a friend of the company.[30] Originally, Mosconi thought in terms of a total state-owned oil monopoly, but after a few years as head of YPF he decided that a jointly owned, public-private monopoly scheme would be the best solution for YPF. Nationalists, especially those on the left, often overlook this latter aspect of Mosconi's thought and try to portray him as supporting a pure state monopoly. This interpretation hardly does justice to Mosconi, for although his hostility to the foreign oil companies was unrelenting, he was also a pragmatic man who recognized the bureaucratic and financial weaknesses of state companies and who believed that a mixed public-private monopoly would combine the twin goals of efficiency and nationalism. (He insisted, however, that the private financing be Argentine, not foreign.)

With characteristic dedication, as soon as he was appointed director-general of YPF in 1922, Mosconi plunged into study of the company's problems. He knew that YPF needed a complete reorganization as well as an expansion program to enable it to compete successfully with the foreign oil companies. As a result, in 1923 the new director-general launched an ambitious development plan that aimed to increase YPF's crude output fourfold between 1924 and 1927 and to make the state oil company a vertically integrated operation by the end of the decade.

Mosconi enjoyed President Alvear's support for this development plan. Like Sáenz Peña, Alvear believed in free enterprise and supported the traditional pro-British export orientation of the Argentine economy. But also like Sáenz Peña, Alvear was no great friend of the United States, and he strongly opposed the American oil companies. He was able to adopt a nationalist oil policy without alarming his British friends, for British petroleum companies had shown little interest in exploration and production within Argentina and were instead primarily interested in the lucrative oil import trade. It was the Americans who were moving aggressively into Argentine oil production, and the British would discreetly applaud Alvear's campaign against these Yankee competitors. Alvear quickly demonstrated his commitment to petroleum nationalism when he agreed to a request from Mosconi to grant YPF virtually total administrative automy. This decree, in 1923, empowered the president to appoint YPF's director-general and its board of directors, but it then gave the state company operating independence except for ministerial approval of its annual budget.[31]

Now that he had authority to make decisions and to implement them quickly, Mosconi moved ahead with his development plan. To increase output, he ordered a dramatic expansion in drilling at the Comodoro Rivadavia fields. The increased profits that resulted from additional production would finance major improvements. During the 1920s, Mosconi proposed and YPF built a string of new facilities at Comodoro Rivadavia, including docks, waterworks, repair shops, and a modern electric power plant. New ocean-going tankers joined YPF's fleet. But Mosconi's principal project, what he called "the most solid pillar of our new organization," was a large oil refinery at La Plata, near Buenos Aires. Built by Bethlehem Steel of the United States, the refinery was financed by a government loan to YPF of 25 million pesos (about $11 million U.S.). YPF repaid the loan by the late 1930s.[32]

The refinery enabled YPF to expand its product range dramatically. By the end of the decade, the state company marketed kerosene, gasoline, aviation fuel, diesel fuel, and lubricants. While the refinery was under construction, Mosconi turned to another important area: creation of a distribution and sales network to market YPF's new products. By 1928, the state company possessed a national network of storage depots and a sales network with over twelve hundred outlets.[33] Throughout Argentina, in the lucrative Buenos Aires market as well as in remote rural areas that the foreign oil companies had avoided because they were not very profitable, YPF's gas pumps, painted the blue and white national colors, reminded Argentines that Mosconi had kept his promise of creating a vertically integrated state oil company.

YPF grew impressively under Mosconi's direction, but its development was more rapid in the areas of plant modernization and

Table 1: Argentine Petroleum Production and Imports, 1922–30 (Thousands of Cubic Meters)

	Total Petroleum Consumption	Domestic Production and % of Total Consumption[a]			Imports[b]	
		YPF	Private Companies	Total Domestic	Quantity	% of Consumption
1922	1495	349 23.3%	106 7.1%	455 30.4%	1040	69.6
1923	1720	407 23.7	123 7.1	530 30.8	1190	69.2
1924	2031	554 27.3	187 9.2	741 36.5	1290	63.5
1925	1802	624 34.6	328 18.2	952 52.8	850	47.2
1926	2348	744 31.7	504 21.5	1248 53.2	1100	46.8
1927	2772	823 29.7	549 19.8	1372 49.5	1400	50.5
1928	3142	861 27.4	581 18.5	1442 45.9	1700	54.1
1929	3393	872 25.7	621 18.3	1493 44.0	1900	56.0
1930	3431	828 24.1	603 17.6	1431 41.7	2000	58.3

[a]Crude petroleum. [b]All petroleum products.

Sources: "Resumen estadístico de la economía argentina," Revista de Economía Argentina 20 (Nov. 1938): 323, 328; Adolfo Dorfman, Evolución industrial argentina (Buenos Aires, 1942), p. 143.

systems expansion than in the fundamental area of crude oil production. In fact, YPF's crude output lagged far behind the ambitious goals Mosconi had set in his four-year plan (see table 1). The cause was a lack of sufficient exploratory drilling, and this in turn reflected the state company's serious shortage of investment capital. During the 1920s, the number of producing wells rose, but production per well dropped—an indication that the original oil fields were becoming exhausted. As early as 1924, geologists warned Mosconi that a vastly expanded program of exploratory drilling was essential to reverse this production trend. Mosconi, however, rejected this advice on the grounds that YPF's financial position was not strong enough to finance a large-scale exploration program.[34]

In his official reports, Mosconi emphasized that YPF's profits ratio was healthy, but it was not large in oil industry terms, and it was insufficient to finance both the massive plant modernization of the 1920s and a greatly enhanced exploration program. As table 2 shows, sales quadrupled between 1923 and 1930, but profits did not keep pace. In an industry requiring huge amounts of money to drill the exploratory wells needed to expand production rapidly, YPF was something of a pauper. And no financial help was forthcoming from the government. Alvear may have backed Mosconi's nationalistic goals, but the president was a fiscal conservative and was unwilling to finance exploration with budget funds.

The profits problem was indeed serious, because it meant that YPF was not able to keep production up with Argentina's rapidly increasing demand for petroleum. The Argentine market for oil products was by far the largest in Latin America, and it was rapidly growing. In fact, by 1920, Argentine petroleum consumption exceeded that of France. This demand pattern was largely due to automobile fever, which swept Argentina during the twenties; by 1929 the republic ranked seventh in the world in motor vehicles registered (with 310,885—more than double the number registered in Brazil, which ranked second in Latin America).[35] As table 1 shows, petroleum consumption more than doubled during the 1922–30 period, but YPF's ability to provision this expanding market fell steadily after 1920. As a result, petroleum imports almost doubled during the 1926–30 period.

Faced with these production and consumption trends, Mosconi made a momentous decision: to give his highest priority to achieving YPF control over *all* Argentine oil production. The director-general believed that if the state company were to survive over the long run, it would have to enter oil production in the interior provinces of Salta, Jujuy, and Mendoza, and it would have to limit the foreign oil companies, which were busy acquiring concessions in these provinces, to areas in which they were already producing oil. YPF, in other words, would have to become a nationwide producing oil company. This policy would require YPF to undertake expensive exploration and production programs in these provinces, postponing plans to expand exploration in Patagonia. Mosconi's highest priority was not to make Argentina self-sufficient in petroleum, at least not in the short run, but to make YPF a virtual national monopoly. Ironically, while Mosconi justified this policy in the name of nationalism, Argentina remained heavily dependent on oil imports—a condition that lasted as late as 1960 and eventually placed a severe burden on Argentina's foreign exchange position.

YPF and the Provincial Rights Issue

Mosconi was deeply concerned about the activities of the foreign oil companies, which had entered Argentina before the war but intensified their efforts during the immediate postwar period. They had acquired numerous exploration permits and some concessions near Comodoro Rivadavia, in Neuquén territory, and in Salta and Mendoza. If YPF were limited to its holdings in Patagonia, Mosconi reasoned, while the foreign oil companies gained control over the rest of Argentina's oil lands, the nation would lose control over oil production and YPF's potential for growth would be limited. To Mosconi, such a scenario was intolerable, since nationalist military ideology regarded government control of the oil industry as vital to national defense. And Mosconi was particularly alarmed because the foreign company that was most actively picking up Argentine oil concessions was Jersey Standard, the American giant whose reputation in Argentina was not exactly unblemished.

President Alvear shared Mosconi's apprehensions about the expansion of the foreign oil companies. In January 1924, he issued a decree that converted much of Patagonia into a state oil reserve; other decrees through 1927 extended this reserve to all the national territories except southern Santa Cruz and Tierra del Fuego. These decrees effectively limited the foreign companies in the territories to the concessions on which they already were producing oil.[36] Alvear's decrees ensured YPF's access to Patagonian oil resources, but another situation in Salta province troubled Mosconi greatly. Argentina's 1853 constitution as well as the prevailing mining code (Argentina as yet had no petroleum code) restricted the national government's powers over mineral concessions to the national territories (which included all of Patagonia). This provision enabled Alvear to issue his 1924 reserve decree, but the central government lacked the authority to administer mining or oil concessions in the provinces, a power that the constitution left to provincial governments. In Salta, where high-quality petroleum was known to exist, the provincial government had conceded much of the province's best oil lands to foreign companies, primarily to subsidiaries of Jersey Standard. Salta was one of Argentina's poorest provinces; it had little other than the sugar industry and a small cattle export trade to Chile to sustain its population. Petroleum thus promised Salta a kind of economic salvation, but the oligarchy that dominated Salta politics preferred to deal with the American oil companies rather than YPF. This situation can only be understood in the light of a century of hostility between the Argentine interior provinces and Buenos Aires. From the standpoint of Salta, the national government had sacrificed the traditional industries of the Argentine interior on the altar of free trade while offering nothing in return. In a similar fashion, if YPF were to control Salta's oil, the greedy porteños could be expected to use the petroleum revenues to support a bloated bureaucracy in Buenos Aires while Salta would receive only crumbs. The salteño elite concluded that it was preferable to deal with the foreign oil companies, which offered to pay production royalties, than with the porteños who managed YPF.

Rivalry between YPF and Standard for Salta's petroleum thus brought the thorny issue of provincial rights to the forefront of Argentine oil politics. Mosconi regarded Salta's position as an outrage and as a virtual declaration of war against YPF. He feared that if Salta's position prevailed, YPF would be excluded from what he believed was one of Argentina's richest oil regions and Standard consequently would acquire a commanding role in Argentine petroleum production.

Mosconi was so convinced of the need to establish a national YPF monopoly and to exclude Standard from Salta that he played fast and loose with scientific advice. On the staff of YPF was the German-born Anselmo Windhausen, who would become perhaps the most distinguished Argentine geologist of his generation. Wind-hausen was convinced that Salta was a much less promising oil region than Patagonia, which, he wrote, held vast untapped oil re-sources "that will be the foundation of the nation's economic life for many centuries." Windhausen urged Mosconi to concentrate YPF's resources on Patagonian exploration, but this advice did not suit the director-general's political goal.[37] Mosconi rejected Wind-hausen and sought alternative scientific opinions. They came from what YPF insiders called the "Italian gang," a faction of YPF's geo-logical staff that was deeply hostile to the German scientists around Windhausen. Led by Enrico Fossa-Mancini, the Italians scoffed at Windhausen's evidence, claimed that the structural complexity of the Comodoro Rivadavia region made exploration prohibitively ex-pensive, and argued that Salta was a more promising area, particu-larly because its oil was of higher quality than Patagonia crude. Mosconi sided with the "Italian gang" in 1923, and as a result Windhausen resigned from YPF to begin a career at the University of Córdoba. In 1926, Mosconi named Fossa-Mancini as YPF's chief geologist.[38] Some insight into the problems that YPF's scientists faced when dealing with the director-general comes from the com-mentary of Erwin Kittl, a member of the German group who was still alive in 1981. Kittl recalled that "Mosconi wanted to direct the geologists like an army, but he had insufficient geological knowl-edge . . . he wanted to impose himself on the geologists."[39]

By 1926, Mosconi had reached a scientific verdict that dovetailed nicely with his political goals. He told a congressional committee that Salta was "the richest oil zone in the country" and that YPF must gain access to it.[40] This decision to challenge the province's petroleum rights had momentous consequences for YPF. It not only diverted the state company's resources from exploration and pro-duction in Patagonia; it also led to a bitter political and constitu-tional struggle that contributed to the overthrow of President Yri-goyen in 1930 and to the end of Mosconi's career in YPF.

Mosconi's objective was to convince Congress to enact a pe-troleum code that would give the national government power to administer petroleum concessions in the provinces. President Al-vear agreed with Mosconi on this issue, and so did the Yrigoyenist wing of the Radical Party, which had split away from Alvear in 1924 in a conflict largely over Yrigoyen's domineering leadership. After 1924, in fact, there were two separate Radical parties. Ex-President

Yrigoyen was preparing for another bid at the presidency in 1928, and he quickly saw the appeal of the petroleum nationalism issue among Argentina's middle and lower classes. YPF, for one thing, was becoming an important source of employment for the large and underemployed middle class. And, at least as significant, YPF was establishing an image as friend of the consumer.

The Yrigoyenist Radicals (called the Personalists) moved quickly on the congressional front. In 1927, with the support of most other parties but against the bitter protests of the interior provinces, they passed legislation in the Chamber of Deputies to establish national jurisdiction over oil resources. A week later, by a much closer vote, the Personalists pushed through the Chamber another law establishing a national oil monopoly. This latter legislation went further than Alvear wanted (and Mosconi also had reservations about it). The interior provinces, led by the oil producers, cursed this policy, as did the urban-based Socialists, who claimed that a state oil monopoly would be a huge fount of corruption and bureaucracy that would enable the president (they were really referring to Yrigoyen) to create a powerful political machine. But the Socialist Party was also suffering from a schism over personalities and leadership, and the breakaway group, which called itself the Partido Socialista Independiente (PSI), backed the oil monopoly. The PSI had just enough seats in the Chamber to provide the Personalists with a margin of victory in the oil monopoly vote.[41]

But victory in the Chamber of Deputies did not mean success for Yrigogen's monopoly plan. Although the aged Radical leader (he now was 76) was reelected president in 1928 and although his party gained a majority in the Chamber of Deputies, the Personalists remained a distinct minority in the Senate, where each province held two seats and where the interior provinces, which were small in population, had much greater weight than in the lower house. Thus, it was not surprising that the senators refused to act on the oil issue. Indeed, they almost totally ignored it, infuriating the president and his party, which launched a newspaper campaign attacking the Senate.

During its 1928 sessions, the Personalist-controlled Chamber of Deputies took Yrigoyen's petroleum nationalism still further when it passed legislation to expropriate the property of the private oil companies and to grant YPF a complete monopoly over Argentine petroleum production. No one in the Radical Party seemed sure how the oil companies would be paid for their properties or indeed whether they would be compensated at all.[42] The vagueness of this legislation reflected the haste with which it was prepared and the obvious political motivation behind Yrigoyen's oil policy. Indeed, the principal political thrust of Yrigoyen's second presidency was a combination of populist and anti-*yanqui* appeals to the urban masses, by now a clear majority of the Argentine population. The president's objective was to mobilize public opinion behind the Radical Party in order to take control of the Senate by means of elections in some provinces and by means of presidential intervention by others.[43] Appalied at the increasingly demagogic character of Yrigoyen's politics and alarmed at the possible consequences, influential figures among the upper class began to discuss deposing the president.[44]

While Argentine politics neared the point of breakdown, General Mosconi attempted to find a compromise to the thorny oil policy issue. His insistence on national control over provincial oil concessions infuriated the Salta elite, and he argued that YPF must have an eventual monopoly over the Argentine oil industry. But, ever pragmatic, he believed that Yrigoyen's plan to expropriate the private companies was financially irresponsible and might invite a trade boycott.[45] It was preferable, Mosconi thought, to limit the foreign companies to their present holdings, which would permit them to exit Argentina gradually. Mosconi also rejected Yrigoyen's plan for a completely state owned oil monopoly, which, his experience managing YPF convinced him, would lead to undercapitalization, excessive bureaucracy, and lagging production. The solution that he suggested in 1929 was a monopoly owned 51 percent by the state and 49 percent by private Argentine investors. This mixed monopoly would be modeled along the lines of the highly successful Anglo-Persian Oil Company, which was 51 percent owned by the British government. And, as with Anglo-Persian, Mosconi thought that the private investors should administer the company, although government members of the board of directors would have veto power over major policy decisions. Argentine capitalists since 1914 had suggested this kind of mixed private-state enterprise to operate the government's Patagonian oil fields. Although domestic investors had not risked much capital in oil exploration, they had expressed interest in participating in the state oil operations once the richness of the Comodoro Rivadavia reserves became clear.

Whether the national capitalists could actually have raised the necessary investments for a mixed monopoly with YPF is, however, unknown.[46] These original and promising ideas for the future of YPF remained untried, for Yrigoyen insisted on pushing towards the total state monopoly, a position that the Senate refused to consider.

Although the political effort to enact a petroleum code and to create a YPF monopoly was stalemated, Mosconi pressed forward on another front to build the image of YPF among the Argentine public. He appealed to the pocketbooks of consumers and voters by dramatically reducing gasoline prices. The price cut, which took place between August and November 1929, amounted to 17 percent. Equally important, in February 1930, Mosconi also established a policy that YPF has maintained ever since, a uniform national price. Before 1930, oil and gasoline prices were higher in the interior areas, which were distant from the coastal refineries, than in Buenos Aires and the other large cities. But after 1930, Argentine consumers paid the same (lower) prices whether they were in remote Catamarca or in downtown Buenos Aires. Faced with YPF's price reductions, the foreign oil companies also lowered their prices.[47] They were deeply concerned, for the Argentine market was large and lucrative, and Mosconi was ready to break any company-inspired oil import boycott by importing petroleum from the USSR, an arrangement that Yrigoyen supported.[48]

The 1929 price cuts were a key event in YPF's history, for they inaugurated a company tradition, which became virtually a political necessity, of maintaining Argentine oil prices below world market levels. In fact, for several decades after 1929, Argentine gasoline sold for a lower price than anywhere else in Latin America except Mexico and Venezuela, and it was among the lowest-priced gasolines in the world.[49] Although YPF's price policy was unquestionably popular among consumers, in later years the state company used advertising to remind the public just who was responsible for Argentina's cheap fuel. "Remember," said one ad of 1934 (when gasoline cost twenty-three cents a litre), "that when gasoline sold for 35 cents a litre, YPF did not exist." Aside from the benefit to consumers, cheap gasoline also propelled new lines of economic activity, particularly the boom in highway transport and road building that took place in the 1930s. And, according to economic historian Adolfo Dorfman, YPF's sales of fuel and diesel oil at low prices significantly fostered the industrialization of Argentina during the 1930s.[50]

But while they benefited consumers, the price cuts were detrimental to the financial position of the Argentine oil industry. For example, the private company Astra, which reported that profits fell in 1930, told its stockholders that despite a rise in international oil prices the preceding year, "the national market got no benefit from the advantage, owing to the sales at low prices effected by the Fiscal Administration."[51] The same was true of YPF. Its profit margin (see table 2), which was already low, fell further in 1930 after the price reductions. One result of this profit squeeze, of course, was that YPF had still less money available for exploration. The evidence, then, suggests that Mosconi cut prices in 1929 not because of economic or financial objectives but to enhance YPF's political position.

But the political tide soon would turn against YPF, for in 1929 and 1930 the Yrigoyen government found itself in a severe and escalating crisis. The Great Depression severely damaged Argentina's foreign trade, the mainspring of its economy, as the value of exports fell 42 percent between 1928 and 1930. The times called for a vigorous and imaginative government to confront the depression, but President Yrigoyen, who was now 78 and appeared senile to many

Table 2: Financial Results of YPF Operations, 1923–30 (in thousands of paper pesos)

	Sales	Expenses	Profits	Profits as % of Sales
1923	16,662	9,662	7,000	42.0
1924	18,279	10,779	7,500	41.0
1925	22,476	12,476	10,000	44.5
1926	32,048	23,048	9,000	28.1
1927	50,141	32,141	18,000	5.9
1928	56,686	41,686	15,000	26.5
1929	58,523	48,523	10,000	17.1
1930	65,871	55,871	10,000	15.2

Sources: Argentina, Ministerio de Agricultura de la Nación, Dirección General de Yacimientos Petrolíferos Fiscales, *Memoria de la Dirección General de Yacimientos Petrolíferos Fiscales correspondiente al año 1928* (Buenos Aires, 1929), pp. 16–20, 36–37; . . . *al año 1930* (Buenos Aires, 1932), p. 26.

observers, was incapable of maintaining the day-to-day operations of the government, to say nothing of formulating a viable economic program. The president, however, was still capable of political vendetta, and Congress was in an uproar in 1929 and 1930 over Yrigoyen's intervention in three provincial governments, which the opposition claimed was motivated by the president's plans to engineer the election of additional Radical senators. Among the upper classes, who had always disdained Yrigoyen's populism and claimed that he was delivering the government into the hands of an uneducated and unwashed mob, the government's paralysis provided a fruitful atmosphere for conspiracy and for making contact with conservative military leaders who long had felt slighted by Yrigoyen and opposed his policies as demagogic. The result was the first successful military coup in Argentina since the 1853 Constitution was adopted. It came on September 3, 1930, swift and almost unopposed, and its leader was Gen. José F. Uriburu, a member of a prominent Salta family. Yrigoyen was exiled. He returned to Argentina in 1932 and died in 1933.[52]

YPF under Attack

General Uriburu, a staunch social and political conservative who opposed the popular suffrage provisions of the Sáenz Peña law, established a dictatorship, closed Congress, and ruled by decree. But the president soon learned that it was easier to overthrow a government than to form a successful new one. Beyond scrapping Yrigoyen's oil policy, repressing the Radical Party, and launching an anti-labor campaign, the new cabinet could not agree on basic policy directions. Uriburu talked of establishing a corporatist order, but he feared the consequences of organizing a mass movement behind the government. He failed to attract support from the agro-exporting elite, which distrusted Uriburu as an incipient economic nationalist. In fact, Uriburu lacked any real base of power, for the army strongman, Gen. Agustín P. Justo, had presidential aspirations of his own, and Justo agreed with the landed elite that Argentina ought to resume its traditional pro-British economic policy and restore constitutional government—although Justo's idea of constitutional rule included the use of political fraud to win elections and to keep the elites securely in power. During 1931, Uriburu recognized that he could not remain in power and called national elections for November of that year. The Yrigoyenist (Personalist) Radicals were declared ineligible to participate, on the grounds of allegedly conspiring to overthrow the Uriburu government, and with the nation's largest party out of the way, the Conservatives and Independent Socialists (who had abandoned their earlier support for an oil monopoly) formed the Federación Nacional Democrática, which nominated Justo for the presidency. Ex-President Alvear, who had hoped to reunify the Radicals, was exiled, and with their leader gone, most of the Alvear Radicals (called Antipersonalists) joined the Justo alliance. Thus was born the Concordancia, the confederation of Conservatives, Independent Socialists, and Antipersonalists that was to rule Argentina for the rest of the 1930s. Justo and a new congress took office in February 1932, and the Concordancia, which Alain Rouquié calls a "sacred union against the return of the plebeians," came into power.[53]

These were indeed difficult times for YPF. Although no direct evidence supports the charge, often made by Yrigoyen's supporters, that the oil companies had conspired with the army in the 1930 coup, Uriburu's cabinet did include several ministers with close business ties to the private oil companies.[54] And this cabinet quickly reversed the petroleum policies of the preceding regime. Mosconi was dismissed only days after the coup and sent into exile. (He returned in 1932, but his health was broken and he died in 1940.) Until March 1932, when Justo named a professional engineer, Ricardo Silveyra, as YPF's new director-general, the state oil company foundered under the incompetent leaders that Uriburu appointed. YPF's financial position reached crisis proportions. The outlook seemed so bleak that during 1931 rumors circulated widely that the government was planning to dismantle the state company. In a hopeful tone, the *Oil and Gas Journal* speculated that YPF's function would become "merely supervisory and regulatory" and that Argentina would again be opened wide to foreign oil investment. Enrique Zimmerman, who briefly served as YPF's director-general under Uriburu, fed these rumors when he stated that he would cooperate with the foreign oil companies and that he wanted new foreign investment in the petroleum sector.[55]

But the Uriburu government was in power long enough to formulate only one consistent petroleum policy, which was to support a production contract between Salta, the president's home province, and Standard Oil. The American multinational's Argentine subsidiary had obtained promising oil concessions from the provincial government in the mid-1920s and was beginning to produce oil when a Personalist Radical, Julio Cornejo, defied the Salta elite and was elected governor in 1928. Cornejo moved swiftly against Standard and ordered it to cease operations on lands that YPF also claimed as the result of a grant of property a private citizen had made to the state company. Standard sued the province and the case went to the Argentine Supreme Court, which in a preliminary judgment ruled in favor of Standard in June 1930.[56] After his coup later that year, Uriburu intervened in the province and a few months later named a personal friend as provisional governor. Standard meanwhile resumed its operations actively in Salta. The provisional government began contract negotiations with Standard, and on November 6, 1931, two days before the national elections, the company and the province signed a production contract that confirmed Standard's concessions and provided for a 10 percent royalty. (The provisional government also signed a contract with YPF, which, however, did not gain particularly attractive concessions.)[57] The contract between Salta and Standard included a particularly controversial provision allowing the company to construct a pipeline from the Bolivian border to a railhead in Salta. Argentine nationalists were outraged, for they had long charged that Standard intended to use Salta as a bridgehead to import oil from its Bolivian fields, which had no outlet to market, and that, moreover, Standard would use a pipeline in Salta to "dump" its Bolivian oil in Argentina at cheap prices, a move that might damage YPF seriously.[58] In December 1931, President-elect Justo forced Uriburu's cabinet to suspend the Salta contract, but the provincial autonomy issue was far from settled and would return to become a major question in Argentine oil politics during the Justo presidency.

The Concordancia's Petroleum Policies: An Overview

The six-year Justo presidency established petroleum policies that had a decisive and lasting impact on the development of YPF. The government halted the political momentum that had been gathering during the preceding decade to create a YPF monopoly and carved out a secure place in the Argentine oil industry for the private companies, particularly the petroleum importers. Justo, however, did not attempt to dismantle YPF, and his reserve policy guaranteed the

Table 3: Argentine Petroleum Production and Imports, 1930–39 (Thousands of Cubic Meters)

	Total Petroleum Consumption	Domestic Production and % of Total Consumption[a]			Imports[b]	
		YPF	Private Companies	Total Domestic	Quantity	% of Consumption
1930	3431	828 24.1%	603 17.6%	1431 41.7%	2000	58.3
1931	3432	874 25.5	987 28.8	1861 54.2	1571	45.8
1932	3214	902 28.1	1187 36.9	2089 65.0	1125	35.0
1933	3234	922 28.5	1254 38.8	2176 67.3	1058	32.7
1934	3476	836 24.0	1394 40.1	2230 64.1	1246	35.8
1935	3833	944 24.6	1329 34.7	2273 59.3	1560	40.7
1936	4068	1140 28.0	1317 32.4	2457 60.4	1611	39.6
1937	4465	1262 28.3	1338 30.0	2600 58.2	1865	41.8
1938	5060	1430 28.3	1284 25.4	2714 53.6	2346	46.4
1939	5213	1625 31.2	1334 25.6	2959 56.8	2254	43.2

[a]Crude petroleum. [b]All petroleum products.

Sources: "Resumen estadístico de la economía argentina," Revista de Economía Argentina 20 (Nov. 1938): 323, 328; Argentina, Comité Nacional de Geografía, Anuario geográfico argentino, suplemento 1942 (Buenos Aires, 1943), pp. 123–24.

state company a major role in the future oil production. But by stripping YPF of its autonomy and by forcing it to submit to heavy taxation, the Justo government left YPF with such severe administrative and financial weaknesses that it was unable to take full advantage of the oil reserves under its jurisdiction. YPF did grow during the thirties. Its crude output and refining capacity expanded considerably, and it developed new product lines. Nonetheless, the oil policies of the Concordancia enabled YPF to provision the Argentine market only partially, and the nation remained heavily dependent on petroleum imports (see table 3).

Justo's oil policies represented, in political terms, a compromise among the nationalists who supported YPF, the oil-producing provinces, and the foreign oil companies. The modus vivendi reflected the fact that YPF was far from being a top priority of the government. The Justo administration, which entered office near the depth of the depression, was first and foremost committed to cementing a stable economic relationship with Great Britain, and this pro-British economic orientation was to affect YPF in various ways. The emergence of the powerful Imperial Trade Preference movement in Great Britain had deeply alarmed the Argentine agro-exporting elite, which formed one of the chief political bases of the Corcordancia. As a result, the Justo government negotiated the famous Roca-Runciman Pact of 1933, which guaranteed continued access to the British market for Argentine beef and wheat but also gave Britain important advantages in the Argentine import market. The need to placate the British meant that any major move against the foreign oil importers, of which the partially British-owned Royal Dutch–Shell was the second largest, was out of the question. To maintain British confidence in the Argentine economy, the Justo government adopted a highly orthodox financial policy. Argentina, alone among the Latin American countries, did not default on its external debt during the 1930s, and to meet its debt payments at a time when government revenues were low, Justo not only cut expenses but also raised new revenues from a variety of sources, including YPF.[59]

The Justo government at the same time had to face other political demands that affected Argentine petroleum policy. The Concordancia was an alliance of parties from both the pampa and the interior provinces, but its most powerful figure was Robustiano Patrón Costas of Salta, the president of the Senate. The strength of the interior in the government coalition meant that the Justo government would be unable to pursue an aggressively nationalist oil policy in the provinces.

In sum, commitment to the British economic connection and reliance on the Concordancia placed severe limitations on the government's oil policy. As a military man who believed that Argentina had to enlarge its industrial base in the name of national defense, Justo did not attempt to dismantle YPF. And, like Mosconi, Justo adopted a hostile stance towards Standard Oil, the only American oil company of any consequence in Argentina. He made conciliatory gestures towards the large and vocal lobby of petroleum nationalists centered in the Radical Party and also now in the Socialist Party. Because they represented urban consumer opinion, these nationalists were a force to be reckoned with and, indeed, they gained important new support in the late 1920s and in the 1930s from Argentina's fledgling industrial sector, which was beginning to produce supplies for YPF. Mosconi had begun a policy of purchasing materials, whenever possible, from nationally owned industries, and he formed a particularly close relationship with Torcuato Di Tella, an ambitious entrepreneur who headed a machinery firm named S.I.A.M. (Sociedad Industrial Americana de Maquinaría). Di Tella was eager to diversify away from the household appliance business in which S.I.A.M. originally had concentrated, and he found the opportunity he needed when Mosconi purchased the gasoline pumps for YPF's retail sales network from S.I.A.M. YPF's orders enabled the Di Tella firm to expand significantly. Justo continued this impetus to Argentine industrialization via YPF's orders, and the links between YPF and S.I.A.M. remained strong.[60]

Despite his desire to placate the petroleum nationalists, Justo could not move too far in their direction without endangering his position with the Concordancia. As Justo maneuvered among these political and economic forces, the result was an uneasy coexistence with the private oil companies. The following section will analyze the principal oil policy decisions of the Justo era in more detail.

The Concordancia's Petroleum Policy: Decisions and Their Impact

YPF's Finances and Taxes

One major problem that the government immediately faced in 1932 was YPF's finances, for the state company was teetering on the verge of bankruptcy. On paper, YPF's position looked sound, for its 1931 budget estimated income of 92 million pesos (U.S. $28 million) and expenses of 79 million (U.S. $24 million), but these figures were illusory, for in fact millions of pesos of the "income" consisted of accounts receivable from other state enterprises—especially the national railroads—that were long overdue. YPF had been delivering oil products to agencies of the government for nothing since 1929. As a result, YPF faced a cash shortage in 1932 so serious that it could not meet its obligations.[61] When Congress opened in 1932, YPF appealed for relief, and the Justo government, aware that it had to do something or see YPF go under, responded by proposing a bond issue to cover the debt of the government enterprises to YPF. Because of the chaotic state of the government's accounting, no one seemed to be able to determine just how large this debt was. Estimates ranged as high as 25 million pesos ($7.5 million U.S.), but the eventual payment to YPF was 12.2 million. Congress approved the relief legislation, but not without strong opposition from YPF's conservative enemies.[62]

This relief represented only a financial palliative for YPF, however, for President Justo also decreed in 1932 that YPF would have to continue selling its products to government agencies at prices below market levels, a condition that caused a serious financial hardship to the state company and continues to the present.[63] The net contribution YPF made to the government during the 1930s on account of reduced prices reached nearly 21 million pesos (see table 4).

YPF's obligation to provide oil products at reduced prices to the government was only one of several new taxes and financial burdens it was forced to assume during the 1930s. A second major tax began in 1932, when legislation obligated YPF to contribute at least 10 percent of its earnings annually to the federal revenues (the actual amount would be decided each year when Congress determined the budget). The context for this tax was the complaints of the private companies that YPF received privileged treatment as well as the extremely severe fiscal crisis that the government faced. (That same year the government also imposed Argentina's first income tax.)[64] But in any case, the 1932 law imposed a heavy new obligation on the state oil company. In 1933 and 1934, Congress set YPF's contribution at 30 percent of its profits. During the rest of the 1930s, Congress did not fix the contribution as a percentage of profits but

Table 4: YPF Taxes and Other Transfers to National and Provincial
Governments, 1932–39 (in thousands of current value pesos)

	Concept (see key below)						
	(a)	(b)	(c)	(d)	(e)	(f)	(g)
1932	3000	—	—	—	3000	1887	4887
1933	3846	—	210	1068	5124	2925	8049
1934	2502	—	—	2101	4603	3934	8537
1935	5000	1585	142	2221	8948	3000	11948
1936	5000	1984	187	2416	9587	1998	11585
1937	3500	2211	187	4178	10076	1550	11626
1938	3500	4126	260	4334	12220	2792	15012
1939	10000	4170	775	4847	19792	2736	22528
Totals	36348	14076	1761	21165	73350	20822	94172

Key:
(a) Federal tax on earnings
(b) Royalties (federal)
(c) Royalties (provincial)
(d) Provincial sales taxes on
     gasoline absorbed by YPF to
     maintain national uniform
     price
(e) Total taxes paid or absorbed
     (sum of first 4 columns)
(f) Loss of income resulting from
     sales at reduced prices to other
     state agencies
(g) Total amounts transferred from
     YPF to governments and their
     agencies
     Source: Adolfo Silenzi de
     Stagni, El petróleo argentino,
     2nd ed. (Buenos Aires:
     Colección problemas
     nacionales, 1955), p. 153.

instead levied a flat sum, which ranged from 3.5 million to 10 million pesos per year but which in every case was well above the 10 percent minimum.[65]

A third category of YPF's financial obligations to the government began with the oil law of 1935. It imposed a production royalty of 12 percent on all petroleum output by privately or publicly owned companies. This sum was payable to the federal or to the provincial governments, depending on the jurisdiction in which the production took place. Passed over the objections of both YPF and the private companies, the 1935 law aimed to provide the federal govern-

Table 5: Financial Results of YPF Operations, 1934–39[a] (in thousands of current value pesos)

	Sales Income	Expenses[b]	Gross Earnings	Taxes	Net Profit	Net Profit as % of sales	Taxes as % of Gross Earnings
1934	92,361	83,756	8,605	4,603	4,002	4.3	53.5
1935	100,009	87,560	12,449	8,948	3,501	3.5	71.8
1936	105,970	91,380	14,590	9,587	5,003	4.7	65.7
1937	123,473	92,451	31,022	10,076	20,946	17.0	32.4
1938	152,118	124,962	27,156	12,220	14,936	9.8	45.0
1939	167,878	132,284	35,594	19,792	15,802	9.4	55.6

[a]1931–33 financial records were unavailable.

[b]Includes amortization and reserve funds.

Sources: Table 4 and Argentina, Ministerio de Agricultura de la Nación, Dirección General de Yacimientos Petrolíferos Fiscales, *Memoria correspondiente al año 1934* (Buenos Aires, 1935), p. 7; . . . *al año 1935* (Buenos Aires, 1936), p. 99; . . . *al año 1936* (Buenos Aires, 1937), p. 107; . . . *al año 1937* (Buenos Aires, 1938), p. 202; . . . *al año 1938* (Buenos Aires, 1939), p. 133; . . . *al año 1939* (Buenos Aires, 1940), p. 136.

ment a royalty comparable to that in other countries and to satisfy the oil provinces' demands for a share of the profits of the oil industry.[66]

By 1935, the basic tax structure under which YPF since has operated was in place. In addition to the taxes already noted and in order to maintain the uniform national gasoline price, YPF also paid the sales taxes that various provinces levied on gasoline. As table 4 shows, loss of income due to reduced-price sales to the government along with taxes on earnings, royalties, and the provincial sales tax forced YPF to sacrifice 94 million pesos to the state during the 1932–39 period. This transfer of income from YPF to the government was about equal to YPF's entire earnings during the 1920s (see table 2) and drained the state company's profits heavily. Table 5, which summarizes YPF's financial performance for 1934–39, shows that the company's net profit (after taxes) was low—in the 3 to 4 percent range—in the mid-thirties, although profits did improve substantially at the end of the decade. The company's tax burden halved its earnings during this period, as Table 6 shows. In other words, the state's policy of using YPF as a source of revenue drastically reduced the company's earnings—and thus its investment funds.[67]

YPF did earn sufficient profits to expand exploration (with highly productive results in Mendoza province) and to increase its refinery

Table 6: YPF's Taxes as Percentage of Gross Earnings, 1934–39 Average (in thousands of current value pesos)

	Gross Earnings	Taxes	Taxes as % of Gross Earnings
1934	8,605	4,603	53.5
1935	12,449	8,948	71.8
1936	14,590	9,587	65.7
1937	31,022	10,076	32.4
1938	27,156	12,220	45.0
1939	35,594	19,792	55.6
Totals	129,416	65,226	50.4

Source: Table 5.

capacity. Yet the tax burden did not permit it to invest enough to make Argentina oil self-sufficient. The government, wrote Oneto in 1941, had forced YPF into the situation of "the millionaire who dissipates his wealth . . . without thinking of tomorrow."[68] This was an overstatement, for YPF was able to make some new investments. Nonetheless, the state company could not afford to subsidize consumers on the one hand and the government on the other without reducing its rate of investment sharply. This analysis suggests that YPF's rate of growth, which critics have termed too slow, was at least partly the result of political decisions rather than of organizational inefficiency.

YPF's Autonomy

The second principal focus of the Justo government's oil policy was to strip YPF of much of its autonomy and to place it under closer government supervision. This was accomplished by the 1932 oil law, whose other major provision was the earnings tax. The Alvear government had granted YPF ample administrative autonomy in 1923, and Mosconi had used it aggressively to promote the company's expansion. The private oil companies had resented YPF's autonomy, which, they believed, had enabled the state company to review all oil concessions that private parties solicited in Argentina.[69] And, indeed, Mosconi had considered it part of YPF's mission to superintend the private companies zealously. YPF's ample use of its autonomy had made it many enemies, not only among the private companies but also in the oil provinces. The Justo government, however, was determined to end what it considered undue favoritism of YPF. As Minister of Agriculture Antonio de Tomaso said in 1932, YPF "would be only one more operating company, without any kind of privileges." In 1932 Congress, following a proposal of Justo, clearly limited YPF's power to operating oil properties owned by the state. To keep the state firm closely in check, the 1932 law also forced the YPF administration to seek cabinet approval for practically every important investment or operating decision including the purchase of new equipment, the construction of new facilities, the use of credit, the setting of prices, and the signing of contracts with the provinces.[70]

The impact of this legislation was to slow YPF's decision making to a ponderous crawl, as the company had to pass an endless stream of petitions to the cabinet, where officials often delayed action for weeks or months. All too often, bureaucratic delay caused material shortages and forced important projects to come to a halt. The dozens of pages that each of YPF's annual reports devoted to listing cabinet decrees on matters of YPF operations is eloquent testimony to the awkward bureaucratic structure that the Justo government forced on YPF.[71]

Petroleum Concessions and Provincial Autonomy

The government was devising policies that affected the growth and development of YPF fundamentally. In effect, it limited YPF's expansion by taxing away much of its profits and controlled the direction of YPF's growth by keeping its management on a short leash. Nonetheless, Justo by no means desired to eliminate YPF or to relegate it to obscurity. Indeed, his government took several important steps to enhance its position in the Argentine oil industry. Justo's determination to stake out a clear future for YPF is apparent from an analysis of the government's approach to the petroleum concession policy. In pressing for the right of YPF to produce in the provinces, Justo was not threatening the position of his British economic allies, for no British oil companies were interested in Argentine provincial concessions. The only major foreign company that would be affected was Standard Oil of New Jersey.

In Argentina the federal government since 1853 held jurisdiction over mineral and oil lands only in national territories and not in the provinces. This division of authority created major political problems for YPF as well as for the private companies, all of which wanted to produce oil in both the territories and the provinces. The rub was that the federal government, beginning with Alvear's 1924 reserve decree, had prevented the private companies from expanding in the Patagonian territories, while the provincial governments had excluded YPF, although the state company had gained a small foothold in Salta. The bitter dispute over provincial and national petroleum jurisdiction had helped undermine the Yrigoyen government, and the Justo administration recognized that it had to solve this problem permanently. The result was a compromise that satisfied the aspirations of Salta province but also enabled YPF to produce oil nationwide.

In 1931 Standard signed a production contract with Salta, but President-elect Justo, who feared that the American company would gain a monopoly in the northern province, forced the contract's suspension. After Justo took office in 1932, YPF and Standard both lobbied vigorously with the Salta government. But while these discussions, during which some salteños accused YPF of stalling, went on, the Salta government teetered on the edge of bankruptcy. To pay its teachers and other employees, Salta needed oil revenues fast. The provincial government that had been elected after the end of Uriburu's intervention decided that it had to act, and on April 7, 1933, it stunned the nation when it revealed a new production contract with Standard.[72] In Buenos Aires, newspapers and most leading political figures expressed shock and disbelief that Standard was gaining a favored position in what was believed to be Argentina's richest oil province. Headlines of mass circulation newspapers like *Ultima Hora* and *Crítica* shouted that "Standard Oil Incites Rebellion in Salta!" and "Creole Fascists Attempt to Sell Argentine Oil to the Foreign Companies!" Congress was the scene of a nasty debate in which Senators Carlos Serrey and Patrón Costas of Salta, together with a few Conservative allies, faced hostile questioning by numerous supporters and opponents of the Concordancia. Salta's policy became the topic of even a pornographic review in a Buenos Aires burlesque theater, which was appealing to "low plebeian instincts," as Senator Serrey indignantly told his colleagues, by portraying American oilmen bribing the Salta legislature in a most unbecoming fashion.[73]

The Justo government closed the offending theater, but the president had to do more than that to extricate himself from an uncomfortable political situation. He opposed Standard and desired to secure a place for YPF in Salta, but he was unable to intervene in the province and cancel the contract. Justo's hands were temporarily tied and Salta was able to defy the national government primarily because Senator Patrón Costas, who was one of the key figures holding the Concordancia together, threatened to withdraw his support from the government. Patrón Costas criticized the "demagoguery" of Argentine oil politics in no uncertain terms and warned that a provincial intervention would mean a political declaration of war. This was something that Justo could not afford, for during those same months in 1933 his government was engaged in delicate negotiations that were vital to the future of Argentine meat and cereal exports to the United Kingdom. Political crisis in Argentina would have precluded these negotiations, which resulted in the Roca-Runciman Pact, and would have destroyed the Justo's entire economic strategy.[74]

The Salta-Standard contract survived and Standard expanded its operations in the province, but the Justo government counterattacked to prevent Standard from expanding elsewhere in Argentina and to reserve the nation's undiscovered petroleum for YPF. While public opinion focused on the conflict in Salta, YPF was gaining a strong position in Mendoza, a province that as yet produced little oil. Mendoza's economy revolved around its wine industry, which had prospered behind a high tariff wall that national governments had maintained against foreign wines since the nineteenth century. When the depression hit, the wine industry faced a crisis, which Justo exploited by implicitly threatening to reduce government protection. At issue were the province's oil resources, which had sparked a good deal of interest among speculators during the 1920s. In 1932, the Conservative governor of Mendoza moved quickly to make peace with Justo. The governor converted the entire province into an official oil reserve and then signed a contract that gave YPF a virtual production monopoly. (Private companies retained only a few concessions they had held since the 1920s.) In return, YPF agreed to pay an 11 percent royalty (which rose to 12 percent in 1935) and to carry out a major exploration program within one year.[75] This agreement was significant because Mendoza contains very large oil resources that YPF tapped by the end of the 1930s. The rich production of these Mendoza wells was one of the main reasons that YPF's production grew rapidly at the end of the decade (table 3).

The second front on which Justo moved against Standard was in the national territories. As will be recalled, President Alvear had converted most of the territories into a state petroleum reserve in which YPF enjoyed an exploration and production monopoly. The only exceptions allowed, which were few, regarded private concessionaires who were fulfilling completely the strict time limits on exploration and drilling that the mining code stipulated. In 1930, Uriburu, in one of his administration's few acts of support for YPF, added Tierra del Fuego to the state reserve. Then in 1934 Justo reaffirmed the government's determination to protect YPF's future production by decreeing that until Congress passed a petroleum code the entire area of the national territories would remain a government oil reserve.[76] This decree meant that in only the provinces of Salta and Jujuy, along with the Mendoza reserve, could private oil companies realistically hope to obtain new concessions. As table 3 shows, their output began to fall after 1934. Now that the government had limited severely the possibilities of expanding production, the companies began to pull their investment out of Argentine oil production and to increase their oil imports.

The companies did not receive much better treatment when Congress passed Argentina's first national petroleum code in 1935. The 1887 Mining Code, which was not written with oil in mind, had regulated petroleum exploration and production in Argentina. Everyone agreed that the nation needed a petroleum code, but efforts to enact one in the 1920s had become hopelessly embroiled with the federalist and state monopoly issues. In 1932, Justo asked Congress to enact an oil code, but again the divisive issue of provincial rights slowed its passage. The Senate, under the watchful eye of Patrón Costas, quickly passed a code allowing for provincial jurisdiction, but in the Chamber, where the Socialist Party and many Antipersonalist Radicals demanded national jurisdiction, the Concordancia's oil policy encountered stiff opposition. The Socialists had become much warmer supporters of YPF than they had been in the 1920s, and they explained this shift in position with the not entirely convincing argument that in Yrigoyen's hands a YPF monopoly would have invited bureaucratic and political corruption. In fact, the Socialists appear to have recognized that YPF enjoyed strong support among the urban populace. (The schismatic PSI, which had become unfriendly to YPF since entering the Corcordancia, was steadily losing elections and congressional seats.) In 1934 and 1935, the Socialist Party did not back the oil monopoly scheme outright, but it did demand a strictly limited role for the private companies. The Concordancia's schemes for provincial jurisdiction, argued Socialist leader Nicolás Repetto and his colleagues in numerous well-publicized speeches, would give Standard the opportunity to make Argentina an American petroleum fiefdom. In adopting this position, the Socialists were now aligning themselves close to the Radicals, the largest party in the country. After being banned from the 1931 elections, the Yrigoyenist Radicals, charging that the Justo government was illegitimate, had refused to take part in elections or politics. But in 1934 and 1935, ex-President Alvear, who had returned from exile, was busy reuniting the fractured Radical Party. By 1935 Alvear had patched the party back together, end-

ing its political abstention, although some Antipersonalists refused
to enter the reunited party and still backed the Concordancia. The
Radicals remained true to their heritage by strongly supporting YPF
and national concession jurisdiction—and some resumed the call
for a state oil monopoly.[77]

The strength of political sentiment against the Concordancia's oil
policy was apparent during the Chamber's lengthy debates on the
code in 1934 and 1935. Although the Concordancia held the majority of the Chamber's seats, the opposition placed the government
party on the defensive. Concordancia deputies emphasized the constitutional rights of the provinces over their oil resources and added
that Argentina needed to encourage foreign oil investment to reduce costly fuel imports.[78] The Socialists and Radicals, however,
emphasized that the issue was really Argentina's long-range economic independence. The oil code, they exclaimed, would serve
"imperialism," which aimed to destroy YPF. According to Repetto,
Standard would foster the rebellion of Salta against the Argentine
government just as it had done in Bolivia, where the disastrous
Chaco War was then underway.[79] At times the debate generated into
name calling, personal insults, and screaming, but eventually, in
February 1935, the Concordancia had its way and the petroleum
code became law.

Despite the emotionally charged rhetoric of the Chamber's debates, the 1935 law represented a compromise between nationalists
and provincial autonomists and between the foreign companies
and YPF. But in the long run, the compromise benefited YPF more
than the foreign oil producers. The oil code gave the provinces the
right to administer petroleum concessions and to collect a 12 percent production royalty, whether the producer was YPF or a private
company. But, in a clause aimed at Standard, the code severely restricted the provinces' authority to grant pipeline concessions. Nor
did the code open most Argentine oil resources to the private companies. It enabled the president and the provincial governors to establish state monopoly reserves, and Justo quickly decreed that the
state reserve would continue in all territories noted in his 1934 decree. The governor of Mendoza also reaffirmed his province's reserve.[80] That left only Salta and Jujuy as fields for private oil investment—and the companies were uninterested in Jujuy in the
thirties. The principle of provincial jurisdiction had triumphed in
the 1935 oil code, but in fact the vast bulk of Argentina's oil lands

remained open only to YPF. As a result, production by the private companies began to decline. "For a number of years it has been impossible for the private companies to carry on new exploration work. . . . Practically the whole of the country is under reserve and closed to private exploration," was Standard Oil's gloomy assessment of the Argentine situation in 1940. It continued that "the Company had been producing oil only from concessions granted many years ago and this production has already reached its maximum and is on the decline with no prospects of improvement."[81]

Marketing Policy

Some writers have charged that Justo's real aim in 1935 was not so much to protect YPF as to promote the interests of the oil importers.[82] The analysis in this essay lends some support to this interpretation, for the government's tax policies had severely restricted YPF's finances and thus its ability to explore and produce from the vast state reserves. And a series of events in 1936 and 1937 made it clear that the Justo government was willing to guarantee the importers a share of the Argentine gasoline market.

Struggle for control of the market had in fact been a major theme of Argentine oil politics since the late 1920s, when completion of YPF's La Plata refinery enabled the state company to retail a full line of petroleum products in direct competition with Shell and Standard, the two largest foreign companies. Although both produced oil in Argentina, they supplied the market primarily through their importing subsidiaries, Shell's Diadema Argentina and Standard's West India Oil Company, each of which had refineries near Buenos Aires. Both companies concentrated their marketing efforts primarily on the capital city and its environs, which together consumed about 50 percent of all petroleum products in Argentina.[83] Indeed, Shell and Standard were not much interested in markets outside the greater Buenos Aires region because the uniform national price policy in effect since 1929 meant that gasoline sold in many parts of the interior, where the cost of transportation to retail outlets was high, would have to be sold at a loss. Although Standard did have a small refinery at Salta that provisioned part of the northwest, the foreign companies left much of the interior market, particularly the more remote regions, to YPF. But the state company would suffer financially if it were limited to consumers spread over Argentina's vast and sparsely populated provinces. To be profitable, it needed a large share of the lucrative Buenos Aires market, and this fact brought the state company into bitter conflict with Shell and Standard.[84]

Mosconi had recognized this market situation, and before 1930, he had worked with the Buenos Aires municipal authorities, who were Radicals, to favor YPF in the allocation of retail outlets in the capital. As a result, the number of YPF gasoline pumps in Buenos Aires increased fivefold between 1928 and 1930.[85] But YPF encountered a much colder atmosphere in the municipal government after the 1930 revolution. In 1932, Justo appointed Romulo F. Naón, a lawyer who had represented Standard Oil's 1930 Supreme Court case, as mayor of Buenos Aires, and Naón quickly reversed the policy of granting newly available retail concessions to YPF. A long, tumultuous, and bitter conflict between the mayor and the City Council ensued over this issue. (The council was elected and was composed primarily of Socialists—of Radicals after 1935.) In 1936, the council declared that gasoline sales were a public utility and gave YPF a monopoly over the utility, but the mayor vetoed the plan.[86]

Justo would not tolerate any plan to drive the foreign oil companies out of the Buenos Aires market. That kind of action would have seriously damaged British interests in Argentina, and the government's economic strategy revolved around maintaining close economic ties with Great Britain. Thus, Justo eventually agreed to institutionalizing a share of the Argentine oil market for the foreign companies. This market-sharing arrangement was the result of a complex series of events in 1936 and 1937.

The foreign companies forced the marketing issue in 1934 and 1935 by increasing their petroleum imports and by lowering their prices for this imported oil to undercut YPF. In the press and in Congress, protest arose that the importers were "dumping" oil in Argentina to drive YPF out of business.[87] In July 1936, the government responded with a decree that gave YPF a monopoly over oil imports and the right to distribute these imports among the private companies. At first, the foreign companies reacted to the decree with alarm, for it seemed, as the *Oil and Gas Journal* pointed out, that their marketing would be "entirely at the mercy" of YPF. Apparently to dramatize its opposition, Standard a few months later offered to sell its entire Argentine operations to YPF.[88]

The private oil companies need not have worried, for the July 1936 decree was only a subterfuge to placate public opinion. Congress failed to act on the proposal to buy Standard, probably because the Salta strongmen in the Concordancia did not want a YPF monopoly in their province. Meanwhile, the Justo government was negotiating with Great Britain over renewal of the Roca-Runciman Pact, whose three-year term expired in 1936. Justo could not afford to antagonize the British with an attack on Shell's import subsidiary when the dominions and the protectionists in Britain were criticizing the Baldwin government for being too friendly to Argentina. After the talks had dragged on for months, Argentina and Britain renewed the treaty late in November 1936. By this time, secret negotiations were also underway between the Justo government and the petroleum importers.[89]

Early in 1937, the government and the importing companies reached an agreement to end the marketing dispute and to change drastically the terms of the July 1936 decree. A series of decisions, whose terms were not revealed to the public, apportioned approximately half the national gasoline market to Shell and Standard and the other half to YPF and certain small private Argentine companies such as Astra. Like the 1935 oil code, the market-sharing agreement was in fact a compromise, for YPF received about half the market quota for the capital city, and both YPF and the importers benefited from market stability. While the companies agreed to adhere to the official uniform gasoline price, no price controls were set over other petroleum products. YPF retained technical control over all oil imports.[90]

The public knew none of the details of this market-sharing arrangement until September 1938, when Radical Deputy Victor Guillot denounced it on the floor of the Chamber of Deputies and charged that YPF had formed a new "oil trust" with the foreign companies. One result, he added, was that both YPF and the importers were raising petroleum prices (except gasoline) and were thus driving up the cost of living. Guillot's exposé caused a political sensation. Socialist and Radical deputies demanded repeal of the agreements, but the Concordancia still controlled the Senate. Justo's successor, Roberto M. Ortiz, who took office in 1938 as the Concordancia's candidate after defeating Alvear in a fraudulent election, took no action on the issue. Market sharing remained in effect until 1947.[91]

The National Petroleum Council

The Concordancia was unwilling to create an official agency to regulate the private oil producers or to mediate conflicts between con-
tiguous YPF and private concessions. Although the reserve policy
of 1934 and 1935 prevented the foreign companies from acquiring
new petroleum lands in most of Argentina, the companies already
had acquired numerous concessions both in Patagonia and the
provinces and were producing oil from them. But as a group of
Argentine petroleum engineers pointed out, a great deal of waste
and inefficiency surrounded much of this private production. The
maximum productivity of an oil field required its operation as a
unit, something that was impossible when several concessions covered the same field. And the private producers were simply flaring
natural gas from their wells. (YPF began to market bottled natural
gas during the 1930s under the trade name "Supergas." It became
extremely popular with consumers.)[92]

To deal with these problems, the engineers suggested creating a
National Oil Council, a proposal that YPF supported. Interestingly
enough, the Province of Salta also backed the plan. Indeed, in 1938
Senator Serrey of Salta introduced legislation to create such a council. Serrey explained that his objective was to ensure the rational
operation and maximum productivity of each oil field—an important consideration in Salta, where YPF and Standard worked ad-
joining concessions and where the province's royalties would suffer
from inefficient production. In this instance, Socialist Senator Alfredo L. Palacios, who usually was at loggerheads with Salta over
oil policy, supported Serrey, but the National Oil Council idea encountered strong opposition from the private oil companies and
from their friends in the Senate, the most important of whom was
Matías Sánchez Sorondo. This influential senator from Buenos
Aires province and spokesman for laissez-faire capitalism insisted
that the proposed council really would benefit YPF and damage the
private companies. President Ortiz gave the proposal no support,
and in this instance the majority of the Concordancia stood behind
Sánchez Sorondo and against Salta province. The Senate defeated
the petroleum council plan, and under these circumstances, the
Chamber of Deputies would not even consider it.[93]

The fate of the National Oil Council plan in Argentina provides
an instructive example of the differences in political structures and

policy formation in Argentina and Brazil. As the chapter on Brazil in this book makes clear, the Brazilian National Petroleum Council (CNP) that emerged in the late 1930s had much broader powers than the Argentine council proposed in 1938. In Brazil, the Council regulated the oil industry as a public utility and had extensive authority over refinery ownership and location, drilling concessions, and imports as well as regulation of oil field production. The Argentine legislation of 1938 would have limited the council's authority to the latter function, but even that was too much for the private sector to accept. The foreign oil companies were much better entrenched politically in Argentina than in Brazil, and the Argentine private sector agreed with the foreign companies that expansion of the government's regulatory role was undesirable. Another political factor that was different was the military: in the 1930s there was no consensus within the Argentine army over oil policy or YPF. As we have seen, the factionalism that plagued the army and that revolved around personalities and power groups (i.e., Justo vs. Uriburu) extended into the oil policy arena as well. In Brazil army opinion united behind the CNP plan, which President Vargas, who had ruled without Congress since 1937, created by decree. Such a result was impossible in Argentina, where politics were much less centralized and where the president could not create even a weak version of the CNP without the consent of Congress.

Summary and Assessment

This essay has examined the early years of YPF and has focused on the pioneering effort of the company's first leader, General Mosconi, who guided the Argentine state oil company through a remarkable early growth period. Indeed, YPF was functioning as a vertically integrated oil company a decade before any other Latin American country entered the petroleum business. Despite its early growth, YPF did not develop into the nationwide oil monopoly of the type that was emerging elsewhere in Latin America. The state company enjoyed widespread public sympathy, but it received only lukewarm political support from the agro-export oligarchy that controlled the Concordancia. As a result, Argentine oil policy during the 1930s was a series of compromises. The government reserved the bulk of the nation's oil resources for YPF but also imposed a heavy tax burden on the state company and severely curtailed the autonomy of its management. YPF was thus unable to exploit fully its available oil resources. This policy left Argentina dependent on foreign companies to supply much of the nation's consumption. The marketing agreements of 1937 ratified this dependence, although they also guaranteed part of the market for YPF.

YPF's early development was stunted because of two basic forces at work in the Argentine political economy. One of them was the determination of the petroleum provinces, especially Salta, to administer their oil concessions and to prevent the formation of a state petroleum monopoly. The second was the strong set of economic ties that linked Britain and Argentina. As long as keeping the British market open to Argentine beef and grain remained the top priority of Argentine policy makers, a thoroughly nationalistic oil policy was out of the question. Although Argentina ended its economic alliance with Britain after 1943, in one way or another governments since the 1930s have retained a policy of compromise between YPF and the private petroleum sector. The nationalistic oil policy consensus that emerged in Brazil never appeared in Argentina. YPF's supporters, including the labor unions, the bureaucracy, and the Radical Party, remain politically powerful, but so are the state oil company's opponents, who include much of the private sector as well as (in recent years) the dominant military officers. The result has been a truce between YPF on the one hand and the private oil sector on the other. When Presidents Juan Perón (in 1955) and Arturo Frondizi (in 1958) attempted to reduce oil imports by opening Argentina to private petroleum investment, the protests of petroleum nationalists were so vociferous that the political position of both leaders was seriously weakened. They broke the truce, and in Argentina's conflictual political culture they paid a penalty. There is little hope for a resolution of this divided oil policy situation until Argentina finally arrives at some kind of consensus on the basic direction of national economic policy. Although YPF's future remains unclear, in many respects it depends on whether Argentina in the crisis period following the Falklands War returns to a policy of international economic integration or opts for economic nationalism.

Notes

1. The company's name, liberally translated, means "Government Oil Fields."

2. Petrobrás does not have a marketing monopoly, but it does have a production and refining monopoly. For comparative analysis of the oil monopoly situation, see George Philip, *Oil and Politics in Latin America: Nationalist Movements and State Companies* (Cambridge and New York, 1982), p. 482.

3. Frank E. Niering, Jr., "Private Capital Aids Oil Drive," *Petroleum Economist* 46 (Oct. 1979): 409; and Philip, *Oil and Politics*, p. 125.

4. Bernardo F. Grossling, *Latin America's Petroleum Prospects in the Energy Crisis*, U.S. Geological Survey Bulletin 1411 (Washington, D.C., 1975): 30–31; "Oil Find off Tierra del Fuego," *Petroleum Economist* 48 (March 1981): 121.

5. "Argentina Starts to Tap Vast Reserves," *Wall Street Journal*, June 3, 1981, p. 33; Adolfo Silenzi de Stagni, *Soberanía y petróleo: YPF y los contratos de explotación de petróleo* (Buenos Aires, 1978), p. 67; Donald O. Croll, "Oil Nationalism Modified," *Review of the River Plate* 161 (April 29, 1977): 552.

6. "Doubts over Oil Potential of the Falkland Islands," *Petroleum Economist* 49 (May 1982): 166–67.

7. "Argentina Looks Abroad for Off-Shore Oil Expertise," *Latin America Economic Report* 4 (Feb. 6, 1976): 23; "Oil Resources Complicate Problems of Malvinas," *The Times of the Americas*, Jan. 6, 1982, p. 2; *Petroleum Economist* 48 (July 1981): 318.

8. Croll, "Oil Nationalism Modified," p. 551; "Oil on the Move," *Review of the River Plate* 162 (Aug. 10, 1977): 194–95; Silenzi de Stagni, *Soberanía y petróleo*, p. 23. On private Argentine oil investment, see Niering, "Private Capital," p. 410.

9. Andrés M. Wolberg-Stok, "New Rules for Oil Firms Considered," *Buenos Aires Herald*, March 16, 1982.

10. "YPF Debt," *Review of the River Plate* 172 (Oct. 13, 1982): 334.

11. Silenzi de Stagni, *Soberanía y petróleo*, pp. 25–29, 38, 46; Jorge Obón, *92 días en YPF: El porqué de una actitud* (Buenos Aires, 1976), pp. 87–88.

12. One of the leading critics of YPF, Arturo Sábato, who had headed the state oil company during the presidency of Arturo Frondizi (1958–62), made similar observations regarding oil policy in the 1960s. See his *Petróleo: Liberación o dependencia* (Buenos Aires, 1974), pp. 10–16.

13. *Petroleum Economist* 49 (Jan. 1982): 34. Oil production figures are from Ibid. 49 (April 1982): 157, and 49 (May 1982): 212.

14. For reports of these plans see "Argentina: Commitment to Private Sector," *Petroleum Economist* 49 (April 1982): 156; Wolberg-Stak, "New Rules."

15. "Oil Policy," *Review of the River Plate* 172 (Aug. 20, 1982): 151–52.

16. J. B. Bradley, "Argentine Fuel and Power and the International Coal Trade," *Comments on Argentine Trade* 9 (May 1930): 48; Adolfo Dorfman, *Evolución industrial argentina* (Buenos Aires, 1942), p. 142.

17. The first years of the state oil company are examined in Marcos Kaplan, "La primera fase de la política petrolera argentina (1907–1916)," *Desarrollo Económico* 13 (Jan.–March, 1974): 775–810. For government budgetary support, see Arturo Frondizi, *Petróleo y política*, 2nd ed. (Buenos Aires, 1955), p. 68.

18. For details of this period, see Carl E. Solberg, *Oil and Nationalism in Argentina: A History* (Stanford, 1979), pp. 22–75.

19. Bunge presented his arguments in a series of articles in the *Revista de Economía Argentina*, a prestigious journal that he founded in 1918 and edited until his death in 1943. Two notable articles of this series are "Nueva orientación de la política económica argentina," *Revista de Economía Argentina* 6 (June 1921): 449–79; "Desequilibrio económico nacional," ibid. 15 (Oct. 1925): 265–85.

20. For example Luis A. Vicat, "Combustibles y defensa nacional," *Revista Militar* (Buenos Aires) 23 (Sept. 1923): 347–49; Alonso Baldrich, *El petróleo: Su importancia comercial, industrial y militar: Legislación petrolera* (Buenos Aires, 1927), pp. 14–20. See also Alain Rouquié, *Poder militar y sociedad política en la Argentina*, trans. Arturo Iglesias Echegary, 2 vols. (Buenos Aires, 1981–82), 2: 154.

21. José Luis de Imaz, *Los Que Mandan (Those Who Rule)*, trans. Carlos A. Astiz (Albany, N.Y., 1970), p. 71. One of the first writers to point to the military background of Latin American public entrepreneurs was John J. Johnson, *The Military and Society in Latin America* (Stanford, 1964), pp. 130–32. Also see Werner Baer, Richard Newfarmer, and Thomas Trebat, *On State Capitalism in Brazil: Some New Issues and Questions* (Austin, Texas, 1976), pp. 23–24.

22. Ricardo Oneto, *El petróleo argentino y la soberanía nacional* (Buenos Aires, 1929), p. 95.

23. For the antitrust investigation, see Argentina, Cámara de Diputados de la Nación, *Informe de la Comisión Investigadora de los Trusts, septiembre de 1919* (Buenos Aires, 1919), pp. 57–59, 72–73. Huergo is quoted in Horacio N. Casal, *El Petróleo* (Buenos Aires, 1972), p. 20.

24. On the various British and private Argentine oil companies, see George S. Brady, *Argentine Petroleum Industry and Trade* (Washington, D.C., 1923).

25. There are two biographies of Mosconi; both are brief and impressionistic. See Carlos Guevara Labal, *El General Ingeniero Enrique Mosconi: Una vida consagrada a la patria* (Buenos Aires, 1941); and Raúl Larra, *Mosconi: General del petróleo* (Buenos Aires, 1957). Mosconi's papers never have been made available, but his published speeches and writings provide insight into his early career; see the collection in *Dichos y hechos, 1904–1938* (Buenos Aires, 1938), pp. 34, 94–95.

26. *Buenos Aires Standard*, Feb. 7, 1923.

27. Larra, *Mosconi*, p. 80; Severo G. Cáceres Cano, *Apuntes para la historia de un esfuerzo argentino* (Buenos Aires, 1972), p. 28.

28. Ricardo Oneto, *El centinela* . . . (Buenos Aires, 1942), p. 282; Jorge Scalabrini Ortiz, *Petróleo y liberación* (Buenos Aires, 1975), pp. 22–23; Juan Carlos Vedoya, "El pacto y el petróleo," in *Todo es Historia*, vol. 6: *El petróleo nacional* (Buenos Aires, 1976), p. 147.

29. Mosconi, *Dichos y hechos*, pp. 127–29, 136–38, 198, 209–10.

30. Enrique Mosconi, *El petróleo argentino, 1922–1930, y la ruptura de los trusts petrolíferos inglés y norteamericano el 1º de agosto de 1929* (Buenos Aires, 1936), pp. 35–50; "Coronel Enrique Mosconi," *Petróleo y Minas* 2 (Nov. 15, 1922): 6.

31. For Alvear's decree, see *Review of the River Plate* 59 (April 20, 1923): 973.

32. Mosconi, *El petróleo argentino*, pp. 57–62, 82; Argentina, Ministerio de Agricultura de la Nación, Dirección General de Yacimientos Petrolíferos Fiscales, *Presupuesto general del año 1924 y plan financiero para los años 1924–1927* (Buenos Aires, 1924), pp. 32–48, 64–70.

33. Argentina, Yacimientos Petrolíferos Fiscales, *Desarrollo de la industria petrolífera fiscal, 1907–1932* (Buenos Aires, 1932), pp. 365–74.

34. Ibid., p. 213; Traian T. Serghiesco, *Líneas generales sobre los Yacimientos Petrolíferos Fiscales desde el punto de vista técnico-geológico y económico* (Buenos Aires, 1930), pp. 8–13; "El petróleo nacional," *Petróleo y Minas* 9 (April 1, 1929): 8.

35. American Petroleum Institute, Division of Public Relations, *Petroleum Facts and Figures* (Baltimore, 1929), pp. 20–21.

36. *Review of the River Plate* 61 (Jan. 18, 1924): 171; Ibid. 61 (April 25, 1924): 1005; "Historia gráfica de las reservas argentinas: Leyes y decretos," *Boletín de Informaciones Petroleras*, 2nd epoch, 16 (March 1934): 13–20.

37. For the Windhausen quotation, see his article, "Apuntes sobre la zona petrolífera de la Patagonia meridional," in *La Época*, Oct. 18, 1920. He presented a technical analysis of his research in "Cambios en el concepto de las condiciones geológicas del yacimiento petrolífero de Comodoro Rivadavia (Comunicación preliminar)," *Boletín de la Academia Nacional de Ciencias de Córdoba* 27 (1923): 1–8. I would like to thank his grandson, Rodolfo Windhausen, for calling my attention to these publications and to the scientific controversy that Mosconi confronted.

38. Enrico Fossa-Mancini, "Faults in Comodoro Rivadavia Oil Field, Argentina," *Bulletin of the American Association of Petroleum Geologists* 16 (June 1932): 561, 575; Enrico Fossa-Mancini, "Geophysics as an Aid in the Search for Oil-Bearing Structures in the Argentine," in *World Petroleum Congress* . . . *July 19th–25th, 1933: Proceedings*, 2 vols. (London, 1934), 1: 177–83.

39. Erwin Kittl, letter to Rodolfo Windhausen, May 16, 1981.

40. Argentina, Cámara de Diputados de la Nación, *Sesiones ordinarias 1926*, Comisión de Industrias y Comercio, *Orden del día Num. 66*, p. 56.

41. For details of these political events, see Solberg, *Oil and Nationalism*, pp. 116–27.

42. Argentina, Cámara de Diputados de la Nación, *Diario de sesiones de la Cámara de Diputados* 4 (Sept. 17, 1928): 357–89. (Hereafter this source will be cited as *Diputados*.)

43. The 1853 constitution gave the president (or Congress, when it was in session) authority to intervene in the provinces to correct legal or political abuses by provincial governments. The president would appoint an intervenor who was responsible for running a provisional government and for arranging new elections. Various presidents abused the intervention clause to serve their own political purposes.

44. Rouquié, *Poder militar,* 1: 205–10.

45. Oneto, *El petróleo argentino,* p. xv (letter from Mosconi to Oneto).

46. Ibid., p. xvi. Also see "El monopolio de petróleo," *Petróleo y Minas* 8 (April 1, 1928): 17; and Alfredo L. Palacios, *Petróleo, monopolios y latifundios,* 2nd ed. (Buenos Aires, 1957), p. 100. On private support for a mixed private-state company, see *Diputados* (June 15, 1914), 1: 777–80; Luis Colombo, *El petróleo argentino y la necesidad de su legislación* (Buenos Aires, 1927), pp. 4, 15.

47. "Foreign Producers in Argentina Facing Uncertain Government Policy," *Oil Weekly* 55 (Dec. 13, 1929): 73; Laura Randall, *An Economic History of Argentina in the Twentieth Century* (New York, 1978), p. 199.

48. Frondizi, *Petróleo y política,* p. 251.

49. Adolfo Silenzi de Stagni, *El petróleo argentino,* 2nd. ed. (Buenos Aires, 1955), pp. 55–56. José J. Díaz Goitia, *La riqueza petrolífera argentina en peligro . . .* (Buenos Aires, 1936), pp. 151–52, presents comparative statistics on gasoline prices for major Latin American cities.

50. This advertisement is found in a pamphlet distributed by YPF, *La nación propulsándose por sí misma: YPF organo de impulsión* (Buenos Aires, 1934). Also see Randall, *An Economic History,* p. 199; Dorfman, *Evolución industrial,* pp. 178–79.

51. *Review of the River Plate* 68 (May 2, 1930): 9, 23.

52. Rouquié, *Poder militar,* 1: 184–198.

53. Ibid., 248. The preceding discussion of political events is based on Rouquié's excellent analysis, pp. 220–51.

54. Carlos A. Mayo, Osvaldo R. Andino, and Fernando García Molina, *Diplomacia, política y petróleo en la Argentina (1927–1930)* (Buenos Aires, 1976), p. 191; Frederick Alexander Hollander, "Oligarchy and the Politics of Petroleum in Argentina: The Case of the Salta Oligarchy and Standard Oil, 1918–1933" (Ph.D. diss., Dept. of History, University of California, Los Angeles, 1976), p. 481.

55. "Abandon Oil Monopoly," *Oil and Gas Journal* 29 (April 16, 1931): 143. For Zimmerman's views, see *Petroleum World* 28 (Aug. 1931): 298. Also see Gilbert P. Moore, "Oil Developments in Argentina in 1930," *Petroleum World* 28 (March 1931): 89–92.

Navy Captain Felipe Fliess, who had administered the Comodoro

Rivadavia fields in 1917–21, served as director-general between September 1930 and March 1931. He was replaced by Enrique Zimmerman, who served until Silveyra's nomination.

56. The court's decision became final in 1932. For details of these events, see Solberg, *Oil and Nationalism*, pp. 139–40.

57. For excellent analysis of Salta oil politics during this period, see Hollander, "Oligarchy and the Politics of Petroleum," pp. 485–506.

58. Mosconi, *El petróleo argentino*, pp. 194–95.

59. Peter Alhadeff, "The Economic Formulas of the 1930s: A Reassessment" (MS, Latin American Centre, Oxford University, 1981), pp. 12–13, 20–22.

60. Thomas C. Cochran and Ruben E. Reina, *Entrepreneurship in Argentine Culture: Torcuato Di Tella and S.I.A.M.* (Philadelphia, 1962), pp. 54–59, 80.

61. Díaz Goitia, *La riqueza petrolífera*, pp. 94, 333.

62. *Diputados* 3 (July 15, 1932): 855; *Petroleum World* 30 (Jan. 1933): 18; Argentina, Cámara de Senadores de la Nación, *Diario de sesiones de la Cámara de Senadores* 1 (July 21, 1932): 853–63, 992. (Hereafter this source will be cited as *Senadores*.)

63. Silenzi de Stagni, *El petróleo argentino*, p. 153.

64. Alhadeff, "The Economic Formulas," pp. 2–13. For the legislation (Law 11,688), see *Senadores* 2 (1932): 930–32. The law also obligated YPF to set aside 5 percent of its profits for merit and seniority pay increases to the work force.

65. *Petroleum World* 30 (March 1933): 96; Silenzi de Stagni, *El petróleo argentino*, p. 92.

66. For the legislation, see *Senadores* 2 (1935): 949–54.

67. For discussion of the legislation on petroleum taxes, see Antonio B. Toledo, *El régimen legal del petróleo: (Ley Nº 12.161): Antecedentes, comentario, doctrina y jurisprudencia* (Buenos Aires, 1939), p. 240.

68. Oneto, *El centinela*, p. 16.

69. "Government Restrictions Check Oil Operations in Argentina," *Oil and Gas Journal* 29 (Dec. 11, 1930): 114; Toledo, *Régimen*, pp. 74–75.

70. De Tomaso's statement is reported in *Senadores* 2 (Sept. 20, 1938): 253.

71. Silenzi de Stagni, *El petróleo argentino*, p. 80.

72. Hollander, "Oligarchy and the Politics of Petroleum," pp. 542–79.

73. For examples of critical newspaper reports, see Augusto Bunge, *La guerra del petróleo en la Argentina . . .* (Buenos Aires, 1933), pp. 16–17. Discussion in Congress is in *Senadores* 1 (May 5, 1933): 75–76, and 1 (May 11, 1933): 87–127. Serrey's complaint is on p. 94.

74. Hollander, "Oligarchy and the Politics of Petroleum," pp. 594–650; *Senadores* 1 (May 11, 1933): 119.

75. Frondizi, *Petróleo y política*, pp. 377–78; Toledo, *El régimen*, pp. 426–

27; Hollander, "Oligarchy and the Politics of Petroleum," p. 574.

76. "Historia gráfica de las reservas argentinas," pp. 13–25.

77. On the reunification of the Radicals, see Rouquié, Poder militar 1: 271–72. For the Socialists' explanation of their position, see Diputados 5 (Sept. 23–24, 1933): 252, and 5 (Sept. 13, 1934): 454.

78. Diputados 5 (Sept. 11, 1934): 282–93.

79. Diputados 5 (Sept. 12, 1934): 406, and 9 (Feb. 27, 1935): 35–36.

80. For these decrees, see "Historia gráfica de las reservas," pp. 28–29.

81. For Standard Oil's memorandum, see U.S., Department of State, Records of the Department of State, 835.6363/532, Nov. 5, 1940.

82. For example, Eduardo I. Rumbo, Petróleo y vasallaje: Carne de vaca y carnero contra carbón más petróleo (Buenos Aires, 1957), pp. 91–92.

83. The capital city and the Province of Buenos Aires together accounted for 59.1 percent of total national consumption in 1930 (Carlos Guevara Labal, El petróleo y sus derivados en la estadística [Buenos Aires, 1932], p. 91).

84. For a review of the market situation, see Great Britain, Department of Overseas Trade, Economic Conditions in the Argentine Republic, June 1937 (London, 1937), pp. 16–17; Review of the River Plate 82 (April 30, 1937): 9.

85. Argentina, Desarrollo de la industria petrolífera fiscal, p. 379.

86. On the struggle between the mayor and the council, see Adolfo Dickmann, Héctor Iñigo Carrera, and Adolfo Rubenstein, En defensa del petróleo nacional y por la dignidad de la función pública (Buenos Aires, 1932). Also see Petroleum World 30 (May, 1933): 128; "Argentina," Oil and Gas Journal 35 (Dec. 31, 1936): 75.

87. Diputados 9 (Feb. 27, 1935): 35–36; "Prohibición de la exportación de petróleo y fiscalización de la importación," Boletín de Informaciones Petroleras 13 (July 1936): 10–14.

88. Review of the River Plate 81 (July 24, 1936): 9; H. Stanley Norman, "Argentina Decree Controlling Exports and Imports Viewed with Distrust," Oil and Gas Journal 35 (Aug. 27, 1936): 30.

89. Review of the River Plate 82 (Feb. 7, 1937): 5, 7; Vedoya, "El pacto y el petróleo," pp. 143–50.

90. Frondizi, Petróleo y política, pp. 358–60; Julio V. González, Nacionalización del petróleo (Buenos Aires, 1947), pp. 52–53.

91. Diputados 5 (Sept. 20–21, 1938): 231–32, and 4 (Sept. 8, 1939): 56–67.

92. Guillermo Hileman and Enrique P. Cánepa, El petróleo argentino en la economía, en la doctrina y en la legislación: Exigencias técnicas que plantea la explotación racional del petróleo . . . (Buenos Aires, 1939), esp. pp. 49–50.

93. Senadores 1 (Aug. 25, 1938): 733–49, and 2 (Sept. 20, 1938), 250–54.

John D. Wirth

Setting the Brazilian
Agenda, 1936–53

State control of the Brazilian petroleum industry was assured with the creation of Petrobrás in 1953. State regulation had been established more than a decade before, in 1938, when the exploration, refining, and marketing of petroleum products were decreed a state utility under the National Petroleum Council (CNP).[1] Government ownership of all phases of the industry has never been completed, however, and the concept of monopoly remains controversial. Exploration, refining, and marketing were open to Brazilian private initiative until 1953; petrochemicals have been developed by Petrobrás in joint ventures with private firms; foreign companies have been active all along in marketing; and since 1978 exploration risk contracts have been made available to foreign and domestic private capital under progressively more liberal terms. To date, the state oil company has not been permitted to control national energy policy. Attempts by Petrobrás to run the alcohol program and set energy policy in the crisis months after the second oil shock in 1979 were no more successful than the CNP's efforts in 1939 to bring coal and alcohol production under the mantle of its fledgling bureaucracy. In short, the Brazilian state is dominant in petroleum, but private capital has always been an important factor in the energy equation, and energy policy making as a whole has never been dominated by the oil monopoly.

One major theme in the history of Brazilian petroleum policy since 1936 is the pragmatic and changing search for balance between state capital, foreign multinational investment, and participation by the Brazilian private sector. On the eve of World War II, for example, Brazilian capitalists were active in exploration as well as marketing and refining, and the major foreign oil companies

3. Horta Barbosa being shown the La Teja refinery by ANCAP's Ing. Fratelli, Montevideo, 1939. Courtesy Arquivo Horta Barbosa.

wanted to invest. In the early years, this industry could have evolved as a classic example of *tripé* (tripartite) development. Peter Evans has shown how the *tripé* approach was used in several other industries.[2] But this was not to occur for oil. Nationalist forces gained the policy momentum after World War II and, in the course of a national debate over oil policy in the second Vargas presidency (1950–54), developed a broad-based coalition to sustain an industry dominated by the state. Despite ups and downs in the fortunes of Petrobrás itself, a consensus emerged in favor of state-led petroleum development. And this consensus carried through the many political storms and system changes from Getúlio Vargas' suicide to the present day.

To be sure, consensus is nurtured by the large public relations staff at Petrobrás, with its substantial advertising budget and ample access to the Brazilian media. But consensus rests upon assumptions and beliefs that are deeply held and were shaped during the formative years discussed in this chapter. It centers on the view that the oil industry is vital to national security, a concept that regards petroleum-based energy as a key to economic growth and as a military and strategic resource. Given this fundamental security dimension, it is not surprising that the Brazilian military was involved to a much greater extent than their counterparts in Argentina, Mexico, and Venezuela. In fact, the first important studies of energy and security were begun in the mid-1930s by army engineers, and the CNP from its inception was headed by a military man. From the beginning, these officers assigned a large role to the Brazilian state. A corollary to this security premise is the view that the Brazilian private sector lacked the will, determination, and capital resources to develop the domestic oil industry. As for the foreign oil companies, at best they were regarded as unreliable—because they were unwilling to invest large sums in exploration—and at worst as alien entities that manipulated national development goals for their own limited objective, which was to sell oil products in a rapidly growing market. The final element in this consensus is that the state is adequate to the task, and that Brazil is in step with history. Since Argentina established the world's first state oil company, the trend has been for governments everywhere to regulate and usually to own outright the means of petroleum production.[3]

Views contrary to the dominant consensus, but equally rooted in history, also have developed. One current of opinion sees Petrobrás, with its large capitalization, its huge demands for funds, and its tendency to operate outside central budgetary control, as a historical mistake. If only the major foreign oil companies had been let in to explore Brazil's huge and geologically difficult terrain, it is argued with 20–20 hindsight, all the risks and expenses that have led only to disappointing results would have been borne by the foreign corporations. But if oil had been found in large commercial quantities by the majors, they continue, Brazil then could have adopted the Venezuelan solution as a way to derive the full benefits of foreign participation while taking eventual control. Many nationalists, including critics on the left, fault Petrobrás for not providing cheap oil at less than the world market price—the social welfare mission they expected state ownership to provide—and for never completing monopoly control over the entire industry. Critics from the private sector claim that the huge, distant bureaucracy that runs Petrobrás is not fully under public control and that this has made the current energy crisis worse than it needs to be. To them, the giant state company is basically flawed.[4] Years ago, the private sector wanted the government to stay out of refining and marketing altogether. But today, a generation later, Brazilian entrepreneurs would settle for a larger piece of the action. One general argument often heard nowadays, after a generation of military rule, is that the existence of this and other massive state enterprises such as Nuclebrás, Eletrobrás, and the Cia. Vale do Rio Doce hinder the development of a more open and democratic society. Making these state companies accountable to Congress and to a broader public is a major political issue in Brazil today.

Given these objections, how could a durable, national consensus on petroleum policy exist, something that greatly aids the state's ability to allocate capital and achieve coherence? This solidarity is all the more remarkable because on specific issues oil policy has been extremely controversial and conflictual, in addition to being a genuinely popular concern. Brazilians do seem to have an affinity for pulling together on major policy issues and for building coalitions of interests to sustain consensus for policy implementation. The well-known style of decision making from the top down is also a factor. But to the historian, a more satisfactory explanation lies in the way these issues were raised, engaged, and resolved—in short, in the historicity of the oil question.

To date, Petrobrás has not achieved its primary mission, which is to find and process enough domestic petroleum reserves to protect a rapidly industrializing nation from the crippling impact of balance of payments problems or of a strategic interruption in the supply of imported crude oil. While it is not quite fair to say, as the old joke has it, that "Petrobrás is one of the world's great oil companies—from the ground up," it is true that the oil shocks of the 1970s that ended the era of cheap oil caught the giant company unprepared. Growing rapidly from commercial operations in the 1970s, Petrobrás became virtually a state within a state. It built its grandiose home office building in Rio de Janeiro in splendid isolation from other key institutions of government, including the Central Bank, the Foreign Ministry, and the army, all of which moved to Brasília. With its exploration and service contracts abroad; its fleet of supertankers, pipelines, and a domestic sales network; its refineries to generate working capital that in turn exerted pressure to make profitable investments outside of exploration, as for example in service stations and real estate, Petrobrás began to look very much like another profit-oriented multinational. Not long ago, Petrobrás was heralded as "the eighth sister" by its president; being under great pressure to find domestic oil reserves, his goals today are more modest.

When oil was cheap, purely commercial considerations dictated a de-emphasis on the search for expensive domestic crude. Funding for domestic exploration as a share of the budget for all activities, including commercial operations and overseas exploration, declined from the late 1960s until 1978, when there was a long-overdue return to the company's primary mission.[5] What soon became a crash program produced results: domestic production in 1981 was still only 25 percent of daily requirements of some one million barrels a day (b/d), but by December 1982 daily production was over 300,000 b/d and by mid-1984 it rose to 500,000 b/d as overall consumption declined due to rigorous conservation measures.

Success in recapturing its primary and strategic security mission enabled Petrobrás to help Brazil avert a payments crisis in 1980, but energy dependency still threatens the nation. In 1980, Brazil allocated almost 60 percent of its export earnings to pay for imported oil—a sum reaching $10 billion, or the cost of one Itaipú hydro-electric dam project each year. Concurrently, the Iran-Iraq war cut off almost half Brazil's imported crude supplies, and a fuels crisis was averted thanks to supplies sent by Venezuela, Mexico, and Ecuador (among others) despite a very tight market. Having ceased to purchase crude from nearby Venezuela in the 1950s, Petrobrás found itself deeply ensnared with the volatile and aggressively nationalist producers in the Near East. Because of oil, Iraq enjoyed a five-to-one trade surplus with Brazil, and when Braspetro (the foreign exploration subsidiary) brought in a large new field the Iraqis revised the risk contract on terms much less favorable to Brazil. (This field, at Majnoon Island, is currently occupied by Iran.) There would be no cheap long-term supplies from Iraq. The Iranians, for their part, were threatening to supply only refined products to fuel-short Brazil; this hit at a major strength of Petrobrás, which is its capacity to generate working capital through profits from refining imported crude. Trade diversification is a key foreign policy objective, yet these third world oil companies behaved as badly as any Latin American nationalist writing after World War I could have accused Shell or Standard Oil of doing. Having overcome this crisis, Brazil moved to secure more reliable supplies closer to home and in so doing paid more attention to its American backyard. As discussed in the introduction, Brazil moved to a more pan-Latin energy policy, including the projected new trilateral company, Petrolatín.

To meet its energy requirements in the 1980s, debt-laden Brazil is relying heavily on domestic resources. In addition to stepping up exploration for oil the government is promoting a large alcohol program, better use of the nation's low-grade coal deposits, and a shale oil project even as it builds expensive nuclear power plants and invests heavily in hydroelectric power. Not all these energy sources are cost effective: there are doubts about the true opportunity costs of the alcohol program, for example, and the nuclear projects are being cut back. The immediate need, however, is to save foreign exchange and to control energy requirements through conservation and a comprehensive fuels program. In turn, having control over the energy bill means better management of the national accounts during the current debt crisis.

Petrobrás failed in its bid to gain control of alcohol and coal production, and the giant corporation has lost some of the legitimacy as a pioneer state enterprise it once enjoyed. Petrobrás may still be a state within a state, but its president is no longer more important than the minister of mines and energy. Ironically, the company is relatively less dominant in today's climate of austerity than it was during the era of cheap oil. Because Brazil has little oil in the ground, Petrobrás cannot fully meet its original objectives: strategic security and balance of payments protection. The fulfilling of important secondary objectives, however, shows the company to be effective in some of its other operations: it can function as a trading company abroad, as for example in Iraq; it transfers money to the central government, which since the 1974 oil price hike receives excess oil profits; and it does assure petroleum price stability in the internal market. To this extent, Petrobrás is a flexible policy instrument, able to move with the opportunities.

Brazil is not well endowed with large oil deposits, a bitter truth that Petrobrás, after initial high hopes and heavy expenditures on exploration in the 1950s, had to face before transforming itself into a "full service company" in the late 1960s. (Private companies had dominated highly profitable distribution services before.) This marketing strategy was reversed by President Geisel in 1978; the company was enjoined to find domestic oil at a time when high prices meant that many of the small fields being discovered were viable for production. The compelling need for capital and equipment set the stage for the foreign risk contracts, while foreign companies under contract and Pemex engineers transferred offshore drilling skills to the Brazilians. Given the need for more drilling, the terms under which private capital is attracted were liberalized.[6] To be sure, Petrobrás reserved the best blocks to itself, and results from the foreign test wells have been disappointing. At the same time, private Brazilian companies that can drill under contract may soon be allowed into exploration in their own right.

Yet the state remains dominant in the oil industry. Important questions about the balance of ownership and control over national energy resources are being asked again, but the basic guidelines were hammered out in 1936–53 when the industry was first established. The forms of capitalist enterprise most adequate and appropriate to the new era of expensive energy and heavy debt obligations are being discussed. But it is doubtful that unfettered free enterprise will return. Politically and economically this would be improbable—so strong is the basic consensus hammered out and inculcated over the years.

In addition to the military dimension, the Brazilian history is in one other respect strikingly different from the other cases in this volume. Brazil's state-controlled industry was developed almost completely on the basis of imported oil. Structures, policies, and institutions to run the industry were established *before* any significant oil was found. Along with Uruguay, which as will be shown had great influence on early Brazilian thinking about how to begin an industry, Brazil is an anomaly among Latin American nations with important state oil companies. Both Mexico and Venezuela nationalized their industries after foreign companies had found and developed large producing fields. The Argentine government discovered that nation's first field, but YPF (Yacimientos Petrolíferos Fiscales) developed there in an uneasy sharing of production and marketing with foreign companies and a few national firms. Brazil, without oil deposits, borrowed from Argentine theory to implement and hold a more consistent and nationalist position. It followed Uruguayan practice by emphasizing state-owned refineries as the key to getting started without domestic reserves. However, ANCAP (Administración Nacional de Combustibles, Alcohol y Portland) did not develop into a vertically integrated company like Petrobrás.

During the formative years, basic questions were posed and solutions found. What was the proper role of state, of foreign, and of national private capital in a rapidly industrializing nation without, as yet, any oil fields of its own? Who supported state leadership in the fledgling industry? Why were other approaches rejected? What were the roads not taken? After researching these questions in the late 1960s I interpreted this history as a progression toward national control. Given the national consensus that developed it would have been hard to argue otherwise. Yet in interviews with former CNP officials and other participants I was impressed by their attractive blend of nationalism and pragmatic self-confidence. They had formed strong views on the roads not taken, especially on the options (the so-called options, as they would put it) of private capital. Returning to these issues and having done new research, I now conclude that a full account of oil in Brazil, which is a capitalist country, must include more analysis of the private sector and of the pragmatic search for balance and proportion. To these issues the narrative now turns.

The Horta Barbosa Years

State direct investment in oil production, refining, and marketing began several years before the monopoly was created in 1953. After

World War I, the Ministry of Agriculture began what soon became an underfunded and overbureaucratized exploration program. The National Department of Mineral Production (DNPM), a successor agency, was no more successful in finding oil or in fostering a climate to encourage private investment. After taking over the government drilling operations in 1939, the newly established National Petroleum Council brought in Brazil's first small oil field in Bahia. Up-to-date methods were used: the CNP employed the Drilling and Exploration Co. of Houston and contracted the United Geophysical Exploration Co. of Pasadena to carry out the first systematic study of Brazilian sedimentary formations. Unfortunately, it soon became clear that this small Bahian field could not supply the nation's crude oil needs. Plans for a stepped-up exploration effort were thwarted by wartime equipment shortages, as were plans to enlarge the CNP's first refinery in Bahia. After the war, the CNP built the first large refinery near Santos, bought a tanker fleet, and continued the exploration program—all before the creation of Petrobrás ushered in an era of heavy state investment in all these sectors. However modest in light of what came later, these early CNP projects marked the beginnings of an industry dominated by the Brazilian state.

The first CNP undertakings were carried forward in the face of grave financial difficulties, and in fact it was the Volta Redonda steel project, not oil, that held priority in the commitment of national resources and United States government loans on the eve of World War II. The discovery of a major field in Bahia would have changed this, but the CNP had no such luck. Even so, policy questions about the ownership and control of petroleum arose immediately. CNP efforts to regulate the national market, to control national energy resources, and to develop them directly were extremely controversial. National entrepreneurs struggled against all this, as did the foreign oil companies. At heart, the fight to secure what almost all modern nations regard as fundamental economic rights involved conflicting attempts to define the public good.

Starting in the mid-1930s, the driving force for state control of petroleum was the Brazilian army. And within the military, the leading nationalist was Gen. Júlio Caetano Horta Barbosa, a military engineer who headed the CNP from its inception in 1938 until 1943. The starting point for military thinking on the need to develop basic industries such as oil and steel was strategic: in the new era of mechanized warfare, Brazil was totally vulnerable to a cutoff of fuel oil, lubricants and gasoline; its industries would begin to shut down in sixty days. The need to accommodate economic development was also considered. Having recovered from the depression by 1934, the Brazilian economy was growing very rapidly on the basis of import substitution. Brazil had entered the motor age and had the second-largest automobile fleet in South America. Gasoline consumption was nearly 7,000 b/d by 1936 and was growing rapidly along with the demand for fuel oil and other petroleum derivatives. This market could support a national refining industry. The Matarazzo group built a small refinery at Cubatao near São Paulo in 1934, the same year that Brazilians from Rio Grande do Sul and an Argentine group founded the Destilaria Riograndense de Petróleo S.A., which in 1936 with the addition of Uruguayan capital became the Ipiranga Company. Projects by national capitalists and foreign oil companies to build refineries surfaced with increasing frequency, until the onset of World War II cut off capital and equipment. The critical question in army thinking was how to raise the huge sums needed for an adequate exploration effort while retaining national ownership of reserves and control over disposition of the oil.

The majors that controlled distribution wanted freedom to dispose of any oil they might find through their worldwide marketing networks. They preferred to import refined products from Arruba but were prepared to build refineries in Brazil to protect their market share. Brazilian capitalists were drawn primarily to refining and distributing, where sure profits lay. However, both the Guinle group and Roberto Simonsen were considering exploration in addition to refining. Guilherme Guinle financed the first competent geological survey of Bahia, formed a company (the Empresa Nacional de Investigações Geológicas), and had a drilling rig working there by 1939.[7] With its emphasis on national security, the army searched for a model that would provide the financial resources and advanced technology to establish an industry without giving up control.

The army's proposal to establish a National Fuels Department was probably drafted by Horta Barbosa in 1936, when he was director of Military Engineering. Horta thought the army was the best institution to resolve the complex oil problem, a view he had already expressed to the minister of war.[8] The National Fuels proposal is the earliest available evidence of a development project from the general staff to reach Vargas' desk. In this printed document, the army was to be responsible for the development of liquid fuels, the navy would coordinate the coal industry (notably the investigation of Brazilian coal for coking purposes), and the Ministry of Finance would handle financial matters. A school to train petroleum engineers would be set up by the new Fuels Department. It would also regulate and direct petroleum exploration, giving first preference to national firms and then to foreign companies, with 25 percent of any profits going to the state. National or foreign firms could refine imported crude oil under contract, but in such a way as "not to restrain the future development of a national industry." Evidently, the army wanted this National Fuels Department to have broad authority to develop the industry, including state direct investment in exploration and refining.

Nothing came of this first army plan. The staff report for Vargas was sceptical. Whether in peacetime the military should be involved in what was essentially a civilian industry was doubtful, his aide concluded. He also underscored the project's large operating budget (50,000 contos, or about U.S. $4.3 million at the time).[9] The president was not disposed to take on what the private sector could do. In studying the relationship between the state and private industry in the area of national defense, military leaders in general reached the same conclusion.[10] In late 1937, however, the weight of views favoring public ownership soon increased in the presidency when Vargas and the armed forces established the authoritarian Estado Novo by coup d'etat. Under the dictatorship that lasted until 1945, interest groups had no Congress to represent them and struggled to gain access through corporatist channels, while the army gained power.

How the National Petroleum Council legislation was elaborated in secret and then decreed without warning in May 1938 is by now well known. At the army's insistence, private interests, both foreign and national, had no opportunity to make countermoves. The legislation declared petroleum a public utility, a regulated industry. It also authorized the CNP to engage directly in industrial operations, although the civilians who drafted this decree in secret at the Foreign Trade Council (CFCE) did not call for a state monopoly of refining. Julio A. Barbosa Carneiro, a career diplomat who like his cousin Horta Barbosa was an orthodox positivist, wanted to reserve the new industry for national capital while keeping the foreign companies out.[11] Domingos Fleury da Rocha, an engineer who drafted the main legislation, was deeply involved in delicate negotiations with Bolivia to obtain oil from recently expropriated Standard Oil concessions, oil that the Bolivian treaty of February 1938 stipulated could be processed only in a Brazilian government plant. Binational companies, financed with Brazilian private capital, would drill and explore in Bolivia. Horta himself would have preferred a state monopoly along the lines proposed by Argentina's Gen. Enrique Mosconi, that is, direct state investment with some limited role for private capital, which he tolerated but mistrusted. As for foreign oil companies, Brazil barred them from exploration and refining. And to assure compliance, all stockholders in private companies had to be Brazilian-born nationals.

Here, then, is how the CNP was born: a) as an army project to assure national control of exploration and refining; b) as a rapid, pragmatic response to Bolivia, which was willing to establish a Brazilian exploration sphere in exchange for capital, if all oil found there were processed in a Brazilian government refinery; and c) as an ideological child of Mosconi and the Argentines. In fact, the recent establishment of Yacimientos Petrolíferos Fiscales Bolivianos (YPFB) was virtually mandated by Argentina, and YPF seized the opportunity to extend its influence to Brazil as well. The irony is that in late 1938, Argentina's Senate turned down legislation to set up the original CNP model, and Brazil did not receive Bolivian oil until Armand Hammer's Occidental Petroleum Co. finally developed the oil fields much later, in the 1970s.

The 1938 petroleum legislation left room for national capital to enter exploration and refining, but not for foreign direct investment in these sectors. Standard Oil hastily reassembled a used Canadian topping plant in São Paulo in order to claim rights predating the CNP legislation, but this was not allowed to operate. Concessions of the Venezuelan type were totally unacceptable, although nothing in the early legislation excluded foreign drilling firms from exploring and developing Brazilian oil fields under contract, with payment in a share of the production. This meant that (in theory) the CNP could have authorized risk contracts.[12]

The Uruguayan model was also close at hand. In May, President Vargas received a letter from Rodolfo P. Peracca, an Argentine busi nessman with close ties to British capital. Peracca offered to do for Brazil what he had done for Uruguay in 1934: build a government refinery, train a staff, and supply crude oil from independent sources. It was Peracca who arranged for Foster Wheeler of Great Britain to build ANCAP's La Teja refinery; he also supplied the crude oil contracts from Anglo-Ecuadorian and Lobitos. This enabled ANCAP to get started despite the international oil companies, which did all they could to prevent it, and despite English companies in Montevideo that pressured their compatriots not to sign these contracts. YPF, by contrast, provided technical assistance and trained the Uruguayan staff to run La Teja, which commenced operations in 1937.[13] Based on this experience, Peracca put his case for building a Brazilian state refinery to Vargas.

The foreign oil companies, he wrote, have basic interests that will always be alien to the national interest, and the military should not be dependent on international companies that, in the case of conflict, will inevitably take an international policy toward the supplying of vital oil. Nations seek their own oil supplies not only to assure the success of their industry but also to guarantee their political independence. "Thus the search for oil is subordinate to a general political and economic policy, and only secondarily subject to mere commercial considerations." This was pure Mosconi doctrine. A Brazilian state refinery would provide "a happy combination of all the benefits: independence in relation to foreigners, lowering of fuel costs for transport and industries, and very large profits" that could be plowed back into exploration.[14]

What might be called the "doctrine of large profits from refining" was difficult to realize in practice; both YPF and ANCAP transferred profits from operations—much of their working capital—to the national treasury. Eighty percent of ANCAP's profits in 1937, for example, went to general revenues, mostly to support a deficit-plagued railroad.[15] (After 1953, Petrobrás avoided this classic pitfall of the government enterprise by plowing refinery profits back into operations, especially for exploration. Excess profits after the 1974 price rise were transferred to the treasury, still leaving ample profit margins for operations.) Profits from refining were what attracted the state governments of Bahia, Rio de Janeiro, and São Paulo to plan refineries of their own, and by 1939 all had projects for mixed, public-private companies on the drawing boards. In fact,

Carlos R. Vegh Garzón, the energetic general manager of ANCAP, was contracted to design the Bahian plant, just as ANCAP had retained a YPF engineer to build La Teja.[16]

Preoccupied with building a small staff and with bringing all government drilling operations under his authority—both goals required political infighting with established agencies—General Horta Barbosa did not rush to construct a refinery. Only after oil was discovered by the CNP in early 1939 was he ready to examine the refinery question, and he made a two-week inspection tour of ANCAP and YPF facilities. His very public visit to Montevideo gave the Uruguayans a chance to press home the viability of their model. The Buenos Aires visit was low-key, due to tense relations between Brazil and Argentina, but Horta did visit YPF facilities. No direct record of his private talk with General Mosconi exists, but the old Argentine nationalist, now quite sick and out of power, told him to stay out of distribution and instead to make refining the key to controlling prices and the internal market.[17] Returning from the Plata, Horta Barbosa reported on the trip and his thoughts to military leaders and to Vargas, "who agreed with all my suggestions and reiterated his intention to bolster (prestigiar) the CNP." On June 6, 1939, he noted in his diary, "Today I entrusted Dr. Fleury [da Rocha, CNP vice-president] to draft a bill to monopolize refining."[18]

That petroleum was a national resource, not a regional problem for the state governments to resolve, was an important principle for Mosconi, who battled Salta province over this issue. The Mexicans also faced it with Manuel Peláez. Upon returning from Montevideo and Buenos Aires, Horta Barbosa discouraged Bahia from going ahead with its refinery; Rio de Janeiro state and São Paulo withdrew their projects in 1940, thus laying to rest the issue of whether or not profits from refining could be used to bolster regional economies and, of perhaps equal concern to the centralizing Vargas government, the state political machines. São Paulo and Rio Grande do Sul, the two wealthiest states, offered certain facilities to encourage private groups to enter refining. But there the major issue concerned the allocation of tax receipts, especially on gasoline. The plan to establish uniform pricing (imposto único) of petroleum products to facilitate the development of all Brazilian regions and to finance a federal highway system was also Argentine. Only after hard-fought battles with the two states did Horta Barbosa see this important piece of Estado Novo legislation through the decision-making process to a decree law in 1940. The imposto único was thus an important victory over states with the strongest economies. This is all the more impressive when one recalls that the state revenues of rapidly industrializing São Paulo were growing faster than federal revenues during this period.[19]

Drawing on Argentine and Uruguayan practice, Horta Barbosa always maintained that the CNP was establishing the principle of national control that was being followed or at least discussed by most Latin American nations. In several of his memos and reports to the general staff and to Vargas and in conversation with the president he cited this continental trend. His thinking thus contained a genuine pan–Latin American content that doubtless legitimized in his mind and others' the transfer of ideas and models from the Plata. Some of these ideas, like the unified gasoline tax, articulated nicely with Brazilian conditions. Others, such as national regulation of the nacent oil industry, reflected the centralizing tendencies under Vargas and were congruent as well with military thinking. However, Horta Barbosa's increasing emphasis on refining—to control supplies, hence pricing, and to finance exploration, thus obviating foreign capital—ran directly counter to the desire of Brazilian capitalists to enter this business. The general could not win approval for his monopoly over refining, and the foreign oil companies with the encouragement of senior government officials such as Minister of Finance Fernando da Souza Costa and Foreign Minister Oswaldo Aranha kept the entire exclusionary model under attack. By the time Horta left the CNP in frustration in 1943 to take up a field command, it seemed as if the private sector had battled him to a standstill. This focuses attention upon the question of balance between state, private national, and foreign capital, perhaps the most interesting and least understood issue of the Horta years.

Private Capital and the State

Horta was not anticapitalist, although he distrusted the motives of men who would put private gain over priorities of the collectivity. Written in a tiny, precise hand, his diary shows him to have been scrupulously accessible to all the entrepreneurs, domestic or foreign. This accessibility belies his reputation as a narrow statist. Several national companies such as COPEBA and ITATIG received permits to explore for oil in 1939, and the undercapitalized companies

of José Bento Monteiro Lobato (see below) were given several opportunities to conform to the new government regulations. But if
Horta's office door was open, he did not accept bribes or accommodate special interests. Here was a military officer of the old school:
serious, a staunch nationalist, a former pathfinder in Gen. Cândido
Rondon's Indian Service. When officiating at the annual Indian Day
ceremony, Horta Barbosa insisted on moving it from the customary
site in Rio, a statue of Cuautémoc given by the Mexicans, across the
bay to Niteroi at the memorial to Arariboia, Brazil's own Indian
resistance hero. His diary reveals the day-to-day activities of a dedicated public servant, a man whose rectitude was useful to Vargas,
especially when rationing was imposed in 1942. It also shows his
running battle with the private sector over refining.

For security reasons, only refineries run by native-born Brazilians
using the capital of native-born Brazilians could be approved. In
1938 this flushed out Argentine and Uruguayan capitalists who had
founded two small refineries with Brazilian investors on the frontier area between Brazil and Argentina. The company (Ipiranga,
S.A., Companhia Brasileira de Petróleos) planned to satisfy regional
southern Brazil using crude oil from Shell (Ecuador) via Argentina
and Uruguay. This group petitioned Vargas (in Spanish) to exclude
Ipiranga from the new nationalist laws as a special case. "We are
Uruguayans and Argentines who set up the companies to run these
refineries at Uruguayana and Rio Grande, and we can call ourselves
the precursors of the Brazilian petroleum industry." In the spirit of
Latin solidarity they asked to retain their share ownership while
turning over the management to Brazilians. Intelligence reports
reaching Horta Barbosa and the army claimed that Ipiranga was
being used by the Argentine government for information as Argentine officers working as sales agents in Rio Grande do Sul were
gathering data on the region, especially communications routes. (In
a world gearing up for war, this information has an interesting parallel with the Philippines, where Japanese naval officers were re-
portedly gathering data on American installations and taking the
pulse of public opinion while working under cover as waiters, va-
lets, and gardeners.) Ipiranga survived the divestiture to become the
largest Brazilian private refining and distribution company.[20]

The same security rationale was used to block Standard of New
Jersey's race to reassemble a used Canadian topping plant in São
Paulo in order to claim prior rights. A cargo of crude oil destined for

the Standard Oil plant was on the high seas when the secret nationalist legislation banning foreign ownership of refineries was sprung in July. CNP also rejected Atlantic, Anglo-Mexican (Shell), and the Texas Company proposals to mount refineries. Note that in Argentina, the foreign oil companies had owned large and important oil refineries since early in the twentieth century. Not so in Brazil, which imported almost all of its refined products. Thus Brazil was more easily able to nationalize refining because there was no significant foreign sector to challenge.

Several of the Brazilian private refinery plans surfacing in the late 1930s depended on foreign technology and equipment, which was of course allowed, and on foreign equity participation, which was not. In fact, one thrust of the 1938 legislation, at least initially, was to reserve refining for national companies and to prevent foreign companies from absorbing the new ventures.[21] But it was one thing to live in the shadow of a state-regulated industry, as Ipiranga learned to do. It was quite another to compete directly with the state as producer, on terms perceived as unfavorable to private capital.

Nationality laws were not so stringent for the wildcatters: naturalized Brazilians could prospect for oil, but they could not use foreign capital. This excluded long-time foreign residents in Brazil as well as foreign direct investment. Horta Barbosa's small staff investigated the balance sheets of half a dozen Brazilian companies to check for evidence of foreign participation. Two of Monteiro Lobato's enterprises could not pass this test. According to the CNP, Germans held 40 percent of the stock in his Companhia Petróleo do Brasil, in the state of Alagoas. Properties of the Companhia Matogrossense de Petróleo lay close to the Argentine frontier, and internal company documents revealed overtures to Argentine capitalists, including proposals for roadbuilding.[22] CNP investigators were appalled by the weak technical base of some national companies, coupled with such irregularities as watered stock, fictitious assets, and weak titles. Stringent enforcement of the law prevented several small wildcatting companies from operating.

Less well reported in the literature is the existence of reputable national groups that were prepared to prospect and develop oil fields. With its experience in coal mining, electric power, and construction, the Guinle group became interested in the Recôncavo area of Bahia and soon joined forces with Roberto Simonsen, a contractor and banker. In 1937 they formed the Consórcio Nordeste, obtained drilling concessions, and began operations. Simonsen pulled out immediately when the CNP was organized in July 1938, but Guinle continued until the CNP crew brought in the first well and declared the entire zone a national petroleum reserve.[23] Their concessions were revoked. Simonsen protested to Vargas, but Guinle continued an operation in Bahia and later drilled some wells in São Paulo and Paraná. By 1940 the Matarazzo group, which had mounted a small refinery in 1934, was reported to be negotiating with the United Geophysical Company for a survey of properties in the state of São Paulo. This in conjunction with their interest in a much larger refinery.[24] Here lies the heart of the matter.

Evidently these large groups were prepared to link back from refining to exploration as part of their import-substituting plans. Vertically integrated national oil companies could have developed on the eve of World War II, when the investment climate was still attractive. The regulatory climate was not, however, and Guinle himself talked to Horta "about the inconvenience of the CNP getting into refining, because this would kill the stimulation and cooperation of private parties."[25]

Simonsen organized two of the state government refinery projects. A 4,000 b/d installation to supply the Guanabara Bay region was planned by Foster-Wheeler, the American company. Major funding was arranged through Murray-Simonsen, which was a correspondent to Lazard Brothers, bankers for Royal Dutch–Shell. Upon completion of the plant, Murray-Simonsen planned to lease it back to the state, in exchange for a share of the profits. São Paulo's projected 6,000 b/d refinery and pipeline to Santos, also designed by Foster-Wheeler, was backed by the Bank of São Paulo and Simonsen. Together, these two plants would have dominated the center-south market by supplying 70 percent of the nation's gasoline, 22 percent of its diesel and fuel oil, and 41 percent of its kerosene. But the two projects died in late 1940, victims of the war and what their backers considered an insufficient profit margin under the new imposto único.[26]

The new unified tax on petroleum derivatives did raise imposts on domestically produced gasoline and on imported crude oil, but refining would still have been profitable. At first the majors tried to block these state government refinery projects, but their subsequent change of heart regarding Brazilian-operated refineries was such that Horta considered the state plants to be fronts for their plans. Joint ventures were not legally permissible. What national capitalists objected to most strongly was Horta's desire to restrain and contain private initiative.

In July 1939 a draft decree to monopolize refining reached Vargas' desk. A ten-year limit was to be put on all refinery concessions, after which the operations would revert to the state. The document proposed that, in addition to making guaranteed profits under the imposto único that regulated the market throughout Brazil, refiners would receive tax relief on imported crude, but open-ended concessions were out.[27] For its part, the CNP would construct the Guanabara Bay refinery. Until sufficient supplies of domestic crude became available the only sure way to obtain price controls was through "refineries administered directly by the CNP." Citing Mosconi on this, he pointed to Argentine and Uruguayan practice. (It is well worth noting that the ANCAP state monopoly surfaced just as a Uruguayan private group was ready to construct that nation's first refinery, in the early 1930s.) Moreover, he noted in response to the storm of criticism now coming at him that "it is necessary to bear in mind that the petroleum question is [essentially] *political* and not economic." To military requirements economic needs take second place in the scramble for oil, and he cited André Bihel's *Le pétrole et l'état* to underscore the point that for governments everywhere petroleum was a primary concern. While he denied publicly that the planned reversion of refineries to the state would necessarily lead to monopoly—the ten-year limit could be extended, or the state could run them jointly—this undoubtedly is what he wanted.[28]

Establishing an investment climate attractive to private capital was not Horta Barbosa's priority. The wildcatters faced an uphill struggle before the war cut off capital and equipment, but several companies organized. By 1942, however, Cia. COPEBA was paralyzed and the government requisitioned its drilling equipment. (Among the well-connected investors in this company was the foreign minister's brother, Luiz Aranha.) ITATIG withdrew from exploration activities in early 1943. The Sociedade Ltda. Petróleos de Maraú (SULIPEMA) likewise ceased operations in Bahia. Monteiro Lobato's three drilling companies could not meet CNP requirements for adequate capitalization; they, too, went under. Given what we now know about Brazil's petroleum geology, the private companies by themselves could never have solved the nation's crude oil problems. Yet over the years a myth about the supposed unwillingness of private capital to enter this sector developed. The historical record disproves this myth—which is often cited by those who claim that private solutions are inadequate—especially when refining as an import substitution industry linking back to exploration is examined.

By statute, moreover, representatives from the oil industry could not serve directly on the CNP, which was organized along corporatist lines. The council did accept nominations from the ministries and chose two councillors from names submitted by the two industrial federations. But the law was written so that no representative of an oil company, or any individual active in the business for the last five years, could serve. This brand of corporatism did not favor the private sector, as Adhemar de Barros, the São Paulo interventor, learned when he tried to place one of Monteiro Lobato's associates on the council.[29] Instead of gaining influence for the private sector, its representatives on the council were used to legitimize the CNP's investigations of *all* private companies, including Ipiranga, which with its Argentine and Uruguayan shareholders had top priority for security reasons.

Capitalists interested in refining were more high powered than the wildcatters, and a Sindicato de Distiladores was active by 1939. They succeeded in blocking Horta's 1939 plans to monopolize refining and fought him off through several redraftings of the bill until his departure from the council in 1943. Minister of Finance Souza Costa was hostile, and the business-oriented Conselho Técnico de Economia e Finánças (CTEF) under him opposed the bill on the grounds that the state should not preempt areas appropriate for private enterprise.[30] Horta explored various forms of mixed enterprises, with the CNP predominating, but the mixed formula that worked for the Volta Redonda steelworks did not win acceptance for petroleum. Guinle, who was president of the new Cia. Siderúrgica Nacional, served on the CTEF, where he opposed a CNP project to expand its regulatory purview to the coal industry under a larger agency, to be called the Conselho Nacional de Combustíveis.[31] The CNP had quite enough to do already with petroleum, he wrote. Whatever new scheme for coal is devised, he counseled, "it should include representatives of the coal industry and consumers so that decisions can have a practical and efficient character."[32] As for refining itself, this was not a sector where the government belonged.

Frustrated, Horta resolved by December 1940 no longer to discuss private refinery projects. Vargas, who tended to go along with monopoly, told him this should be done on the mixed enterprise model—with federal, state, and private capital, but the government predominating. For his part, Horta asked to have the CNP Council made into a fully independent agency, with control over hiring and industrial operations. "In sum," he responded to CTEF critics, "on principle what is condemnable is not the state as industrial producer, but the bureaucratization of such an enterprise, whether entrusted to the public administration or to a private entity."[33]

While private interests did not lose these skirmishes, the onset of World War II cut off imports of refinery equipment. In any case the nationalist laws would have needed amending to allow the foreign financing on which several of the large national projects were dependent. To the nationalists, Brazilian capitalists were never free of suspicion: that they would inevitably act as fronts for the foreign oil companies was an argument used against them in the postwar oil debates. The upshot is that Ipiranga owned the sole private refinery after Matarazzo interests shut down in São Paulo. Three more private plants were built after the war; all still operate but since 1953 may not expand. Without funds or support, the CNP did not begin a large refinery project until 1949.

The Foreign Option

Although they provided good service at the pump, the foreign oil companies had a terrible image in Brazil, as in most of the industrializing parts of Latin America. In the cities the newly organized middle classes and labor shared with old-style elites a suspicion, if not a visceral dislike, of anything that big, that powerful, and that foreign. Inspired by YPF and in fact coached by Argentines, who sent intelligence on the majors, the CNP believed that the foreign companies would do anything to prevent the creation of a national oil industry. It was in their marketing interest to sell refined products from abroad; they would move into refining if necessary to preempt the rapidly growing domestic market; and if they were allowed into exploration and found oil, it was in their interest, again as marketers, to hold it in reserve while selling Venezuelan crude that was closer to their big markets. This is plausible enough, to judge from the Argentine case. "Instead of developing Argentina's oil reserves as much as possible with a view to achieving a progressive reduction in Argentina's dependence upon imports, the Standard Oil Company had preferred to import much of its crude from Peru and Colombia where costs of production are lower than Argentina."[34] This was the economic risk, as Horta saw it. Bolivia provided a recent case study of the political risks.

In their drawn-out nationalization campaign against Standard of Bolivia, the Bolivians charged that Standard not only refused to support their losing war effort against Paraguay, but also downplayed exploration in the Oriente, especially along the Argentine border, in favor of Venezuelan imports. The press throughout Latin America believed that the oil companies were behind the Chaco War, a conspiracy theory that is false. What Brazilian diplomats knew all along is that Argentina had been pressing Bolivia to nationalize (which happened in 1937), to organize a state company modeled on the YPF (this was the YPFB), and then to export whatever oil might be produced on the old Standard properties (by financing rail and pipeline links to Argentina). Facing defeat by Paraguay—whose war effort was financed by Argentina—and weakened by the war, Bolivia asked Brazil to build a railroad from Corumbá to Santa Cruz de la Sierra and offered access to the oil fields there. In early 1938 the two nations signed a treaty. That railroad imperialism north from the Plata basin was a long-standing Argentine policy was of course known to the Brazilians, who welcomed Bolivia's invitation to balance this pressure. But for strategic reasons Brazil also wanted Bolivian oil, and in fact Fleury da Rocha was negotiating the terms and conditions for access before the CNP was created. To keep Standard from slipping in through the back door, Bolivian nationalists insisted that this oil would have to be refined in a Brazilian state plant. This suited the Argentines, who had blocked an earlier Standard Oil pipeline-railroad-barge project to supply the Buenos Aires market from Bolivia. In these classic buffer-state politics, petroleum was a major new factor.[35]

As a military man, Horta Barbosa saw Standard Oil's less than enthusiastic support for the Bolivian war effort as proof that in a national emergency the foreign oil companies could not be counted on. Their support for private armies in Mexico was another strike against them, as he told the minister of finance.[36] He insisted, furthermore, that the tides of history were running against them in favor of a pan-Latin solution. This point he hammered home in several memos and conversations with senior officials and the President. The companies were an extension of their own governments when it came to the strategic interests of Britain and the United States. He pointed out to Vargas that Washington had enjoined American oil companies to locate foreign reserves after World War I. If this was a legitimate goal for the United States government, he asked, wasn't it just as legitimate for the Latin American governments to safeguard and develop their own resources?[37]

The companies had friends and associates in the Foreign Ministry, the Ministry of Finance, and the powerful civil service agency DASP, which blocked Horta's efforts to control his own personnel policies. Standard Oil personnel throughout Latin America shared information on host governments. The YPF, as mentioned, supplied Horta with intelligence on foreign operatives and their maneuvers within the Brazilian government. Ultimately, it was the desire of the Brazilian army to exclude foreign investment in petroleum that kept the foreign companies out. Three times Standard Oil offered to develop the Brazilian oil industry: in 1940, 1941, and 1942. To stop these initiatives, Horta used all of his prestige and the threat of resignation.

The Standard Oil proposals should be seen within the context of American policy on the eve of World War II, which was to encourage the oil companies to abandon confrontation policies in Latin America. In April 1941 the American embassy reported that "Standard Oil has recently signified to high Brazilian officials its willingness to cooperate with Brazil in the development of oil properties in this country, provided the necessary modifications are made in existing legislation." In September the cabinet decided to modify the mining laws to allow American producers to operate in Brazil. Paul J. Anderson, Standard's Rio representative, left for discussions with the parent company in New Jersey and returned in early October with approval to commit $10 million for initial expenses in oil exploration. To put this in perspective, the entire Standard investment in Argentina, which had a larger market, was estimated by the company at $29.9 million in producing, refining, and marketing as of mid-1940.[38] In sum, this was a serious proposition.

Evidently Standard was ready at first to propose a mixed private

company, but on liberal operating terms that to Horta were tanta-
mount to a concession. This would be a national company in name
only, "placing in the hands of a foreign financial organization . . .
one of the fundamental elements of our economy and of our security."[39] More than once Anderson had asked what the CNP
needed with respect to its planned refinery, and as early as 1939 he
had proposed drilling in partnership with the government.[40] In
1941, seeing the entire thrust of CNP policy under assault and la-
menting that legislation based "on the experience of other South
American nations, principally Argentina and Uruguay" was up for
revision, Horta asked to resign.

Horta argued forcefully that a Brazilian state company using foreign technology and contracted personnel could develop the industry. But while he could veto foreign enterprise, he could not gather
enough political support within the state to finance his own plans.
Vargas skillfully extracted a steelworks from the United States government after an American private company pulled out, but he hesitated to throw the resources of his presidency behind Horta's program, as Luciano Martins has pointed out. Certainly he was
unwilling to let Horta resign, and as Aranha (who favored Standard
Oil's entry) reported, "Vargas somewhat favors the ideas in general
of Horta Barbosa."[41] Martins speculates that Vargas may have entertained discussions with the oil companies to foster a climate of
good will while the critical steel negotiations were in progress. It
seems more likely that the president was prepared to accept the
foreign companies as the best solution to financing the nascent oil
industry, for as Horta noted in his diary after talking with Vargas,
"he seems inclined to allow the competition of foreigners in the
development of petroleum." Underscoring this policy change,
Vargas told Horta that money destined for the CNP under the five-
year plan would be diverted to buy military aircraft.[42] The cabinet
voted to liberalize the mining laws but, in the face of opposition
from the general staff and Horta's resignation, Vargas backed off.

Another Argentine parallel sheds light on these events. In 1946
President Juan Perón began to give serious considerations to proposals for expanding the role of foreign capital in the Argentine
petroleum industry. According to Robert Potash, Perón may well
have used these discussions with Standard Oil as a way to improve
overall relations with the United States and to ease the embargo on
military equipment. Yet it is also clear that Perón, in conversation

with the American ambassador, maintained "that he had no hope of YPF satisfying the fuel needs of the country, that its operations had been unsatisfactory for years, and the best solution lay in arrangements between the government and foreign, preferably American, companies. . . ."[43] Attempts to modify the mining laws on terms the foreign companies would accept were blocked in the Argentine cabinet. Perón had his labor constituency to consider, and Vargas, who pioneered the politics of populism, faced army opposition in 1941. Despite economics, neither president could change policy, and when both tried again in the early 1950s they lost their nationalist support.

In 1942, Horta asked the United States government to finance the refinery and a major exploration effort. The figure he presented to the cabinet was $35 million. With the war on and the chances for private investment now less likely, Vargas told Horta that he was "inclined toward setting up a mixed company along the lines of the Steel Company." Aranha and Souza Costa asked whether there would be such a credit and told the American ambassador that they hoped "it is not (repeat not) true: they do not believe that such an operation would facilitate the finding of oil in Brazil."[44]

It happened that anything beyond maintaining a modest CNP program in Bahia was not seriously entertained in Washington, where oil interests were well entrenched. As Max Thornburg, the State Department's petroleum advisor commented, concerning

the general development of Brazil's oil possibilities, which means the exploration and examination of an area two-thirds the size of the United States, jungle covered and largely unmapped, any project under the direction of the Brazilian government (except in the broadest sense expressed through its petroleum laws) must be regarded as fatuous. This would be an undertaking warranting the combined resources of a group of large oil companies, under whatever terms Brazil chose to make. We have only to look at Mexico and Argentina (neither of which has discovered a single new producing field) [he said incorrectly] or at Brazil's own effort to develop Bahia—which is the equivalent of a small lease in Texas.[45]

Developing even a small industry in Brazil became less important to American strategic planning as the war progressed toward victory. There was little inclination to finance what private companies were prepared to do. And in any case credits on this scale would have angered the Argentines, who wanted drilling equipment in exchange for supplying petroleum products and natural gas to southern Brazil, Paraguay, and Uruguay.

Horta lost ground. CNP authority to determine liquid fuels policy was curtailed as João Alberto Lins de Barros, another military officer, was made coordinator of economic mobilization with overall authority to plan guidelines for exploration, refining, fuels allocation, pricing, and the use of alcohol additives. The CNP was relegated to implementing policy. Sympathetic to foreign investment in the oil industry, Lins de Barros held the policy initiative. The CNP had to enforce unpopular gasoline rationing and allocate scarce fuel oil as Brazil became totally dependent on the United States–sponsored Petroleum Pool, which supplied petroleum to all Latin American nations except Argentina, which refused to join.

The upshot is that the foreign companies came close to entering Brazil, but there would be no tripé development of this basic industry. On security grounds, the army would not let them in. And while opinion was not monolithic, nationalism within the military remained a strong exclusionary force well into the 1970s. The Venezuelan model based on concessions was controversial, although a version of it (the Petroleum Statute) was proposed by President Eurico Dutra after the war without success. Nationalists such as Horta Barbosa argued that Venezuela was the classic Latin American case of a nation reaping few benefits from oil exports while remaining socially and economically backward. Furthermore, the major companies represented an international force that was alien to the development of a true national capitalism led by the state. The Vargas government was split on the wisdom of this course. And what sort of national capitalism was appropriate for Brazil was also very controversial. This final theme in the early oil years was brought out nicely in the test of wills between Horta Barbosa and Monteiro Lobato.

The Trial of Monteiro Lobato

An octopus, grasping the national substance from weak, naive, and credulous native societies, was often used as a cartoon symbol for the foreign oil companies. This was an image well suited to the nationalist opinion taking root after 1918 in the expanding cities of Latin America, with their newspaper-reading middle classes and nascent labor movements. It was Upton Sinclair's classic image, brought into Latin America by the press and authors of a new nationalist genre and manipulated by national capitalists. All this Monteiro Lobato, the well-known writer and publisher, did with great success in his 1936 bestseller *O escándalo do petróleo* (*The Petroleum Scandal*), which is still in print.

From the early 1920s Monteiro Lobato dreamed that petroleum would transform agrarian Brazil into an industrial nation. Obsessed with finding oil, he participated in at least three wildcat companies and wrote extensively on the theme. The 1934 mining law separating subsoil from surface property rights made it more difficult to raise capital. He stridently attacked this clause and the bureaucratized government drilling service. This, he said, would not let Brazilian companies drill because Standard agents, working from the inside, spread false rumors that Brazil had no oil, reported pessimistically on geological formations, and hindered the granting of prospecting permits. The octopus, he trumpeted in *Scandal* and countless newspaper articles, did not want Brazilian companies to find oil. Fleury da Rocha, the technocrat who ran the DNPM, was his special target of attack in public, and in private letters to Vargas and other public figures he used the same hot rhetoric to make his points. When Fleury da Rocha drafted the new CNP regulations, Monteiro Lobato continued his fight to discredit what he saw as yet another opportunity for government meddling and inefficiency.[46]

It is true that the DNPM was bureaucratized. In reporting on errors in the sometimes over-hasty government survey reports, Gen. Francisco Pinto, secretary of the National Security Council, told Vargas that the negative conclusions in these reports had brought on "the gravest difficulty, obstacles and failures for the private initiatives in these [geological] regions that the government technicians had condemned. In fact, this caused a serious conflict between private enterprises and official agencies, a loss of precious time and the retarding of a solution to a problem that is important for us." He urged greater caution in the issuing and use of technical reports, in order not "to rekindle old errors, with the retinue of disagreeable consequences for the government, the interested parties, and the country."[47] It was against this background of muddle and confusion that Horta took great satisfaction in bringing in the first producing well at Lobato in early 1939.

It is also true that the technocrats were very sceptical of Monteiro Lobato's credentials, his companies, and his claims that Brazil was awash with oil if only the wildcatters like himself were free to find it. They suspected—in fact they were certain—that he was using nationalism to hype stock in his companies. Monteiro Lobato's firms were financially and technically weak. His questionable business practices are very clear from the extensive CNP investigations.[48] When the CNP refused to authorize his companies Monteiro Lobato ridiculed Horta in several semiprivate letters. In May 1940 he told Vargas that by suffocating Brazilian private enterprise Horta was doing the work of the Standard oil octopus.[49]

At first Monteiro Lobato had welcomed the military's interest in petroleum, for he saw the army as a promising new market for the oil his companies would find if only they were allowed to drill. He was careful not to attack the army as an institution, but he saw the CNP and the nationalist laws as fundamentally alien to the development of a national capitalism. The long letter to Vargas outlined some telling points: that the mining laws made it hard to raise risk capital; that the laws prohibiting foreigners from owning stock deprived his companies of a critical source of funding; and that the tendency of the CNP was toward monopoly. But by accusing Horta of doing Standard's bidding he began to go too far.

In July he wrote an even stronger letter to Army Chief of Staff Gen. Pedro Aurélio Góes Monteiro. In a long and rather defensive reply Góes assured him that private companies in compliance with the law were authorized to function and that the government was leaving ample room for private enterprise in petroleum.[50] Emboldened and perhaps sensing victory, Monteiro Lobato circulated mimeographed copies of his letter to Góes in São Paulo. He laid plans for a National Petroleum Congress with the editor of the Revista do Petróleo, a struggling journal serving the nascent industry. Together they planned to distribute twenty thousand copies of a special issue of the Revista, to all the military leaders and to federal, state, and municipal authorities. The next step would be to hold a national plebiscite on the laws by organizing a letter-writing campaign to Vargas. In August, Matogrossense shareholders began to send telegrams to Catete Palace, while Monteiro Lobato himself sent a telegram to Horta threatening to expose him as the stooge of Standard Oil unless the CNP authorized his companies.[51]

This, in short, was a campaign based on reaching opinion leaders through the mails. (In 1948, the nationalists used this method to stop President Dutra's liberal Petroleum Statute, then in the Congress. Horta's debates in the Military Club with fellow officer Juarez Távora, who defended foreign investment, were mailed to thousands of opinion leaders.) But it was Monteiro Lobato who was the first to try this classic interest group tactic—in defense of private enterprise!

Political action of this sort was extremely risky in the authoritarian Estado Novo, however, and the security police were soon onto him. (Ironically, the same security police shadowed the nationalist rallies Horta attended in the late 1940s, because Communists had infiltrated the nationalist petroleum campaign).[52] Moreover, Monteiro Lobato was known to have been sympathetic to the right-wing Integralista Party, which was banned in 1938. Always ready to run risks, he thought the German government might provide oil field equipment for his Alagoas company. His travels to Argentina on publishing business were noted by the police. But with this publicity campaign and the threat to expose "Horta Cibola" ("Garden Variety Horta") as Standard's stooge, he had gone too far. In late 1940, Horta brought this would-be Colonel Drake before the National Security Tribunal on charges of "crimes against the popular economy."

A national security dimension added drama to this trial. General Horta belonged to the inner group of military officers who strengthened the army's power within the state following the 1932 São Paulo regional revolt.[53] Having commanded a regiment in politically sensitive São Paulo after 1932, he would return again in mid-1943 as commander of the critical Second Military Region, where he oversaw the mobilization of units for the expeditionary force to Italy. His archive contains an extensive commercial clipping file on the oil issue, beginning in 1936 at the highpoint of Monteiro Lobato's attacks on Fleury (who became Horta's vice-president at the CNP). This indicates that the politics of petroleum being orchestrated by Monteiro Lobato and his allies was a concern well before the general himself came under attack. And now Monteiro Lobato was publicly impugning his own integrity.

Yet Horta went to great pains to document the case against Monteiro Lobato's methods of operation as a slipshod entrepreneur. He was determined to expose the sharp-tongued, indeed reckless, publisher and author as a fraud. In this regard the trial has considerable interest beyond its role as a civil liberties case during the Estado Novo. For his part Monteiro Lobato relished the fight, even though he exhausted his health and resources. For wanting to destroy this bureaucratic meddler in uniform he was sentenced to six months in jail.

It was a battle of sharply contrasting personalities. The General, a serious and careful man, had family roots deep in the coffee zone of Minas Gerais, known for its nationalist traditions and regional manufacturing. Years of soldiering on the frontier with Rondon and the Indian Protection Service and with army construction battalions had given him a national vision of Brazilian development problems. As a practicing orthodox positivist, he was prepared to seek out the best in any man but to come down hard on error.[54]

Monteiro Lobato, by contrast, was the classic outsider from Taubaté, with its ultraconservative, hidebound coffee elite in the Paraíba valley. In novels, short stories, and children's tales he wrote brilliantly for the emerging middle class audience in São Paulo, and more than anyone else he developed the first mass market for Brazilian books through his influential publishing house. Under his leadership, the Companhia Editora Nacional commissioned distinguished works on Brazilian history and society. Infatuated with American industrialists—he served briefly as consul in New York and sketched a biography of Henry Ford—Monteiro Lobato was not very careful in his business dealings. But his rationale was passionate: it was nothing less than to energize Brazilian society with a new elite of risk-taking entrepreneurs, through the oil industry.[55]

The trial was a battle to define and defend the public good, a battle of perceptions, a sort of Estado Novo morality play in which definitions of national capitalism were at issue. On the cutting edge of state-sponsored industrialization, Horta Barbosa was part of the emerging state apparatus that was organizing the nation for rapid economic growth based on heavy industry. The aim was both to make Brazil less vulnerable in the event of war and to create the infrastructural building blocks of an industrial society. On the cutting edge of national capitalism, Monteiro Lobato saw himself as the champion of an American-style development model based on the free play of factors, which the state should facilitate but not organize. Denying the international oil companies access to the national subsoil wealth was good. Restricting national companies from necessary access to foreign capital, to large profits from refining, and to land titles for raising risk capital was all detrimental. To him, the entire regulatory thrust of the CNP was wrong: there could be no such thing as a nurturing, or tutelatory, state bureaucracy.

Years of fighting the DNPM and later the army-backed CNP convinced him that industrial decisions made in secret, from inside the state, were a pathology. The CNP thrust was toward state monopoly. This was a fundamental misallocation of resources and a stifling of initiative, Monteiro Lobato wrote Vargas and his ministers several times.

Philosophically, this was a battle rooted in the heritage of Iberia, from which Brazil had received no legacy of the bourgeois revolution. Horta tapped the tradition of state paternalism, with an organic definition of society and the public good. Modernization would be carried out under the aegis of the state. Monteiro Lobato sought a whole new social order based on individualism, rooted in the free market. Because there was so much to sweep away he was virulent. For his part, Horta insisted that authority be upheld, that attacks on government agencies and "the honor of the Council" be redressed. Both men were convinced of their rectitude.

But Monteiro Lobato was broken and discredited, at least in the eyes of official Brazil. Technocrats do not suffer lightly public criticism. Whether in or out of uniform, men of the emerging state apparatus in this industry disliked sensationalism, publicity hounds, amateurs, and wildcatters. Fascination with this trial and the issues it raised lingers on. Juarez Távora, for example, who as minister of agriculture in 1934 had clashed with the wildcatters, had second thoughts. Monteiro Lobato thought Juarez Távora was impressed by the charges of bureaucratic meddling and stifling private initiative leveled in the letter to Vargas that began the government case against him.[56] After the war it was Juarez who became the leading spokesman in the military for a liberal policy on oil. (Years later, when the old marshal reviewed my book for translation by the Getúlio Vargas Foundation, the only change he wanted was a softening of my judgment of Monteiro Lobato as a fraud.) In 1982 Brazil issued a stamp to commemorate the centenary of Monteiro Lobato, not as publisher (where his sure reputation lies) but as petroleum pioneer. He is today a hero of *paulista* private enterprise. Just before he died, however, Monteiro Lobato recanted and came out strongly for state monopoly.

The trial itself was held behind closed doors, and while he regretted taking on the state, when asked whether he actually believed that the Standard Oil octopus had infiltrated the Vargas government Monteiro Lobato, always the polemicist, said definitely yes. Sentenced to prison in February 1941, Monteiro Lobato appealed and was absolved in April on grounds that as a private citizen, writing a private letter to President Vargas, he had every right to express his opinions in what could be considered a personal attack only on the CNP. Freed, he sent Horta a note thanking him for the time in jail, which was put to good use reading a book entitled *A Short Introduction to the History of Human Stupidity*, after which nothing the CNP did could surprise him. Infuriated, Horta interceded with Vargas, and in May Monteiro Lobato was recommitted to prison. He had been in protective custody since March, when his request for a passport was turned down lest he flee to Argentina.[57]

Conclusion

By the early 1940s it was by no means certain that there was going to be a very large and active state role in the Brazilian petroleum industry. General Horta Barbosa and his associates set the agenda for what was to come, although in their time the rate and direction of change were still unclear.

The entrepreneurial capitalism of Monteiro Lobato not only was discredited but also was considered unsuited to the development of Brazil's petroleum deposits. His own shaky ventures could never have brought in the industry. But it bears repeating that at least four private exploration companies were approved by the CNP and began to drill. These enterprises were aborted by the war, and the transfer of Bahian oil-bearing deposits to the status of a national petroleum reserve stifled initiative, as in Argentina. The conventional explanation that the private sector lacked capital, technology, and experience to prospect for oil gained force over the years as the very difficult and problematic petroleum geology of Brazil became better known. Yet it is also clear that regulations imposed by the CNP hindered the rapid development of a privately owned and operated refining industry and that the linking back from refining to exploration by private companies was never really tried. Under Horta, the industry was shown to be politically risky and highly conditioned by the state.

Horta's strategy of development based on a state monopoly of refining, as pioneered by Uruguay, was not adopted until Petrobrás was created in 1953. He could not finance a state refining industry because he lacked the support of major elements in the state, which welcomed multinational capital. The initial model of Petrobrás as a flexible holding company owed much to discussions with the major oil companies. However, the second Vargas government (1951–54) decided that the foreign companies were not prepared to offer enough capital to develop the industry rapidly, and then Congress restructured Petrobrás as a monopoly. In any case the security arguments against the majors had been greatly reinforced in a postwar nationalist campaign (1948–49). General Horta participated, and most of the officer corps was convinced to drop its brief flirtation with economic liberalism with respect to oil.

The strategic arguments of the Brazilian military owed much to Argentina, but Great Britain, France, and the United States also developed arguments in favor of a state-owned petroleum corporation. The British, largely for strategic reasons, invested government funds in the Anglo-Persian Company, which was taken by Mosconi as his model for the mixed corporation. In France the government dominated the industry after World War I. In the United States, the same rationale was advanced in support of the United States Oil Corporation by Senator Phelan, but his ideas carried little weight: American hostility toward state capitalism proved too strong. Had the United States military had more influence, however, the American pattern might have more closely resembled the Brazilian.[58]

The Brazilian army was not monolithic on the nationalist question. The celebrated postwar debates between Horta and Juarez Távora in the Military Club were prefigured within the first Vargas government. Juarez' warnings against creating an overbureaucratized state oil entity carried weight within the military, especially in the immediate postwar period before they were caught up in the oil nationalization campaign. Ever since the army pushed the development of basic industry, especially after 1937, oil as a strategic industry has been important to military thinking. The CNP began under the aegis of the military, and with only two exceptions it has always been headed by a retired army officer.

Horta's biggest victory was to push through the imposto único over the objections of powerful state governments and the private sector. But by 1942 the CNP's innovating days were over. As mentioned, the council was not allowed to become a super energy agency with authority over coal. In October 1941 a National Fuel and Lubricants Commission was established under the National Security Council to coordinate distribution as part of the United

States program to supply the Western Hemisphere. Horta went along, reluctantly. (Argentina did not join the American-run Petroleum Pool, but Uruguay did.) In November 1942, in the face of eroding political and military backing for his policies, Horta lost overall policy control of the petroleum industry to João Alberto Lins de Barros, the new coordinator of economic mobilization.

Alliance with the Americans undoubtedly strengthened the proponents of an internationalist development policy. On the eve of Horta's departure from the CNP, Lins de Barros was seeking to modify the mining code, ensuring better access to subsoil rights, and to create a new Ministry of Mines and Energy that would be receptive to foreign capital.[59] Clearly military circles criticized Horta for failing to achieve the independence he sought through the exclusionary model. Horta cooperated with the Morris Cook Mission, which recommended United States loans and supplies for CNP programs. But on principle he would not accept dependency on the Americans. When Leslie Webb, the embassy petroleum attaché, asked to inspect the Bahian oil fields to ascertain whether the CNP really needed this material, Horta Barbosa pulled him up short. "I told him frankly that tutelage from him or anyone else was inadmissible, that he had authority to verify nothing whatsoever. Either send us the material we wanted or not—but we cannot be put in the situation of colonials."[60]

There were limits to the American relationship, which furthermore did not suffocate Brazilian self-confidence or belief in its capacity to do the job. In Europe, Horta's counterparts in the Romanian state oil monopoly were telling the Germans much the same thing. In fact, there is a fascinating parallel in the way both developing countries were able to escape foreign ownership of their oil deposits. Thus, the Romanians sold oil to the German war economy, but they enmeshed the German companies in nationalistic regulations similar to the Brazilian laws. To monitor the industry more tightly, they established a National Petroleum Council. The two developing countries were on parallel tracks until Romania had the bad luck to be taken over by the Soviet Union, which imposed a different economic system.[61]

Brazil moved into the orbit of American power and vaulted ahead of Argentina, which, despite pan-Latin rhetoric about a common petroleum destiny could not supply Brazilian needs during the war and with which Brazil sustained competing strategic interests in the buffer states of Paraguay and Bolivia. The military balance shifted. YPF had offered to train CNP technicians in 1939, over the strenuous opposition of Standard Oil; the first group went to AN-CAP instead, and the next group to the United States. After the war Argentina became less nationalistic and certainly more divided than Brazil in its approach to the oil industry. The solution adopted by Perón in 1955 was far more compatible with American thinking than was Petrobrás, the nationalist solution of a wartime ally.

Unhappy with the loss of power to Lins de Barros, unable to secure major United States government credits for the refinery and a larger drilling program, and charged with administering a necessary but unpopular rationing program, Horta Barbosa resigned for the last time in mid-1943. Vargas with reluctance let him go.

The government was divided over oil policy, but the partisans of state-led capitalism left the stronger legacy. This was a technocratic view, both military and civilian. Is the swollen, semiclosed bureaucracy that is Petrobrás today the inevitable consequence of this legacy? Horta himself never wanted to create a monolith, and in fact Mosconi warned him against the dangers of bureaucratization and urged him to stay out of distribution, citing the example of political favoritism in the granting of YPF filling stations. On the other hand, Petrobrás is finally (at least in part) fulfilling the strategic mission that was the rationale, from the start, of state intervention in this industry. That this political and economic solution is the better way is firmly rooted in the Brazilian consensus. State leadership began early in petroleum.

Energy policy after the war reflected a more economic, and less strategic, conception of the oil industry, but the earlier framework of strategic, nationalist, and propublic ownership arguments prevailed over liberal internationalist and national capitalist solutions in the open marketplace of ideas and public opinion. Thus, in this democratic era, the second Vargas administration approached the problem of creating a national oil industry primarily out of concern for the balance of payments problem created by postwar growth. (By 1951, oil consumption was over 100,000 b/d and rising 20 percent a year as requirements for gasoline, diesel, fuel oil, and asphalt grew rapidly.) However, the political and juridical form Congress gave Petrobrás, the new state company, grew out of the nationalist structures of Horta Barbosa and his associates, structures on which the durable postwar consensus supporting state enterprise is based.

Before 1953, when Petrobrás was established, Brazil lacked a long-term, autonomous investment policy to support exploration, a tanker fleet, and a refining industry. Unlike the CNP, which from 1938 to 1949 was held to a very tight budget, Petrobrás was designed as a well-capitalized, financially independent company. Imports of refined products fell dramatically in the 1950s, while crude imports for the new refineries went up. For all its resources, Petrobrás today cannot develop the industry alone, and since 1978 space has opened up again for private capital beyond marketing. Finding the right investment balance under national control is again part of the Brazilian agenda as Petrobrás seeks to fulfill its primary mission, which is to secure the nation's oil supply and to protect its balance of payments.

Notes

I wish to thank Col. Luís Augusto Horta Barbosa and his sister, Dona Julice Cardoso, for granting access to their father's papers and a happy place to work. In addition to comments from the other contributors, Frank McCann made very helpful comments on this chapter.

1. The literature is well developed. Consult Gabriel Cohn, *Petróleo e nacionalismo* (São Paulo, 1968); Getúlio Carvalho, *Petrobrás: Do monopólio aos contratos de risco* (Rio de Janeiro, 1976); Luciano Martins, *Pouvoir et développement économique, formation et évolution des structures politiques au brésil* (Paris, 1976); Peter S. Smith, *Oil and Politics in Modern Brazil* (Toronto, 1976); and John D. Wirth, *The Politics of Brazilian Development, 1930–1954* (Stanford, 1970). The most recent account—an excellent survey—is George Philip's *Oil and Politics in Latin America, Nationalist Movements and State Companies* (Cambridge, 1982). For a penetrating commentary on the current energy crisis see Kenneth Paul Erickson, "State Entrepreneurship, Energy Policy, and the Political Order in Brazil," in *Authoritarian Capitalism: Brazil's Contemporary Economic and Political Development,* ed. Thomas C. Bruneau and Philippe Faucher (Boulder, Colo., 1981), pp. 141–76.

2. Peter Evans, *Dependent Development: The Alliance of Multinational, State, and Local Capital in Brazil* (Princeton, 1979).

3. Petrobrás, "Tendências mundiais da indústria do petróleo: Os 26 anos de criação da Petrobrás," in *Cadernos Petrobrás* 2 (Rio de Janeiro, [1979?]).

4. Consult the highly critical study by Alberto Tamer (a reporter for *O Estado do São Paulo*): *Petróleo: O preço da dependência, o Brasil na crise mundial* (Rio de Janeiro, 1980).

5. Philip, pp. 390–91. Chapter 18 summarizes these issues.

6. *The Petroleum Economist* 48, no. 10 (Oct. 1981): 448; *Latin American Weekly Report*, WR-81-40 (Oct. 9, 1981): 7.

7. Geraldo Mendes Barros, *Guilherme Guinle, 1882–1960: Ensaio biográfico* (Rio de Janeiro, 1982), pp. 133f.

8. Memorandum, "O petróleo e a defesa nacional," Col. Horta Barbosa to Minister of War [Eurico Gaspar Dutra], Rio de Janeiro, Jan. 30, 1936, in Brazil, Congresso, *Petróleo, Documentos parlamentares* 2 (Rio de Janeiro, 1957): 6.

9. "Criação do Departamento Nacional de Combustíveis," *Esboço do projecto* (Rio de Janeiro, 1936). This nineteen-page document is in CFCE Processo 480, "Departamento Nacional de Combustíveis," XIII, PR 31, Arquivo Nacional in Rio de Janeiro. Penciled unsigned notes from presidential secretariat, report to Vargas on the DNC plan, in CFCE Processo 480.

10. Stanley E. Hilton, "The Armed Forces and Industrialists in Modern Brazil: The Drive for Military Autonomy (1889–1954), *Hispanic American Historical Review* 62, no. 4 (Nov. 1982): 657. Hilton lays great stress on how the two sectors cooperated to develop an armaments industry, but he does not tell about nationalist elements in the military who clashed with domestic steel producers over steel policy in the late 1930s—at the same time the army and the technocrats were restricting private investment in petroleum.

11. Martins, *Pouvoir et développement économique*, p. 299; Wirth, *Politics of Brazilian Development*, pp. 145f.

12. CNP Ofício, Horta Barbosa to Chief of Staff Góes Monteiro, Aug. 2, 1940, Arquivo Horta Barbosa in Rio de Janeiro.

13. Letter, Rodolfo P. Peracca to Vargas, May 2, 1938, enclosing a memorandum on the refinery proposal, in Document 20787, PR, Lata 38, Pasta 1, Arquivo Nacional. Interview with Eng. Carlos R. Vegh Garzón, Oct. 12, 1982, Montevideo.

14. Peracca letter, ibid.

15. Carl Solberg, this volume, pp. 83–85; Carlos R. Vegh Garzón, "Memorandum sobre las obras y adquisiciones projectadas por la ANCAP para el trimestre 1938–1939–1940," [Dec. 1937], in *Informes Ancap*, Jan.–June 1938, Archivo Vegh Garzón in Montevideo. As a growing enterprise that had not yet fulfilled its mission, he argued, ANCAP should have capital to expand; unlike the Usinas y Teléfonos del Estado, for example, or the National Mortgage Bank, it should not turn over profits to the state. Purchasing a tanker fleet and manufacturing cement, as mandated in the 1931 legislation setting up ANCAP—in sum, completing the company's mission—would bring much more revenues to the nation.

16. Letter, Vegh Garzón to ANCAP President Carlos de Castro, Dec. 21, 1938, in *Informes Ancap*, July–Dec. 1938, pp. 47–67, in Archivo Vegh Garzón. Vegh Garzón also did refinery studies for Paraguay, Peru, and Guatemala in 1938 (interview, Oct. 12, 1982).

17. Wirth, Politics of Brazilian Development, p. 152.

18. "Relatório do General J. C. Horta Barbosa ao Presidente da República, sôbre sua viagem ao Prata, em abril de 1939," in Brazil, Congresso, Petróleo, 2: 279–303. Horta Barbosa, conversation with Vargas on April 20, 1939, and instruction to Fleury given on June 6, 1939, as reported in his private diary (hereafter cited as Diário), March–Oct. 1939, Arquivo Horta Barbosa.

19. Wirth, Politics of Brazilian Development, p. 154; Joseph L. Love, São Paulo in the Brazilian Federation, 1889–1937 (Stanford, 1980), pp. 242–44.

20. Telegram, Aguiar Fernandez Clulow to Vargas, from Rio Grande, RS, July 11, 1938, document 15005, PR, Lata 46, Pasta 3, Arquivo Nacional. Bilhete confidential (carbon copy), unsigned, from Rio de Janeiro, Sept. 16, 1938, Arquivo Horta Barbosa.

21. Martins, Pouvoir et développement économique, pp. 290, 299. Decree Law 4,071 (May 1939) granted the existing and prospective private plants de facto monopoly privileges and guaranteed market zones. The purpose was to protect them from dumping. But Horta Barbosa was already moving to undercut the private refiners with the ten-year term concession, and he wanted a monopoly, sooner or later.

22. "Parecer sobre adaptação da Companhia Petróleo do Brasil ao regime legal," CNP Processo PL 32/38 in Ofício 2553 of Sept. 6, 1939; Ofício 4602, Horta Barbosa to Vargas, Dec. 17, 1940, in the supporting documents for Monteiro Lobato's trial, in Tribunal de Segurança Nacional, Processo 1607 [Apelação no. 756], Arquivo Nacional.

23. Mendes Barros, Guilherme Guinle, pp. 141–42.

24. Voluntary Report (Confidential), commercial attaché to State Department, Rio de Janeiro, Sept. 25, 1940, DS 332.6363/321, U.S. National Archives.

25. Conversation with Guinle, June 8, 1939, in Diário, Arquivo Horta Barbosa.

26. Wirth, Politics of Brazilian Development, pp. 154–55, 249, note 29.

27. The draft decree, along with Horta Barbosa's "Exposição de motivos no. 1.745," of July 17, 1939, is in Anexo 2 of Brazil, CNP, Relatório, 1º Triênio, 1938–41 (Rio de Janeiro, 1941).

28. "Ligeiras considerações sôbre projeto de decreto-lei relativo à indústria da refinação de petróleo" (typewritten), Oct. 18, 1939, Arquivo Horta Barbosa.

29. Adhemar de Barros sponsored Hilário Freyre of the Cia. Matogrossense do Petróleo and the Cia. Petróleo Nacional to be a class representative on the CNP (letter to Vargas, June 20, 1938). A report by DNPM Director Fleury da Rocha on July 18 recommended that the request be denied and expressed his agency's long-held scepticism about these Monteiro Lobato companies, which were still in the organizing stage (Document 560 in PR, Lata 45, Pasta 1, Arquivo Nacional).

30. CNP Ofício, Horta Barbosa to Vargas, Feb. 7, 1940. The CTEF, as Souza

Costa told Horta, "is against industrial development by the state" and maintains that private capital should do the refining, on grounds that any state agencies are inefficient and bureaucratic" (Arquivo Horta Barbosa).

31. In Uruguay, ANCAP's attempt to monopolize coal imports (to control prices) in 1937 was opposed by the importers and distributors and was frustrated by legal uncertainties raised by the Congress. Deeply divided over the wisdom of monopoly, Congress passed the famous Baltar Law in 1936, eroding ANCAP's authority to expand into the distribution of coal, cement, and petroleum. Although it controlled petroleum imports, ANCAP was unable to complete the monopoly by automatically taking over all distribution as soon as its refinery supplied over 50 percent of the nation's fuel needs, a trigger written into the 1931 legislation setting up the company. But when its refinery reached this target in 1937, the Baltar law protecting private enterprise seemed to block the further growth of ANCAP filling stations. The foreign oil companies were equally uncertain of their rights, so in January 1938 a compromise was struck: to protect market shares, the private companies agreed to sell most of ANCAP's refinery output under their own labels, while ANCAP imported the crude oil and did some marketing. This solution lasted for decades and was adopted by Brazil in the mid-1950s, when the large Petrobrás refineries came on stream.

In Argentina, the government did little to develop national coal reserves until the war. YPF had the authority to set fuel prices, but because it did not monopolize refining it had to settle for less favorable terms in dealing with the majors, which were producers and refiners as well as distributors, although YPF did all the importing after 1937. (The majors served only as distributors in Uruguay after 1936 and in Brazil after 1955.) In Brazil, the CNP's 1938 bid to become a national energy agency responsible for both coal and oil was embedded in the larger question of whether to create a Ministry of Mines and Energy. Opposition from the Brazilian private coal interests helped bury this initiative for years. After 1953, the CNP lost power to the new Petrobrás.

32. Mendes Barros, *Guilherme Guinle*, p. 126; Conselho Federal do Comércio Exterior, Ofício 68741, "Melhor aproveitamento do carvão nacional," J. A. Barbosa Carneiro [Horta's cousin, and director of the CFCE] to Vargas, Nov. 12, 1938, and numerous other documents in PR 26563, Lata 50, Pasta 4, Arquivo Nacional. The Transport Ministry initiated this project.

33. Horta diary entry, Dec. 12, 1940, and conversation with Vargas, Dec. 26, both in *Diário*, Arquivo Horta Barbosa. Ofício, "Regulamento da industria de refinação do petróleo e regime tributário dos combustíveis e lubrificantes líquidos minerais," 78-page document, Horta Barbosa to Vargas, Feb. 7, 1940, Arquivo Horta Barbosa.

34. Memorandum of conversation, J. Crane and Peter Leib, treasurer and head of production, respectively, of SONJ with State Department personnel, Dec. 26, 1940, DS 835.6363/532, U.S. National Archives. To be sure, Stan-

dard was unable to produce outside of Salta because of the Justo government's nationalistic reserve policy (consult Solberg's chapter).

35. Based on research in Brazilian diplomatic archives for Bolivia, Paraguay, and Argentina, 1934–39, in Arquivo Histórico do Ministério das Re-
lações Exteriores (MRE), Rio de Janeiro. On the Argentine railroad and
pipeline projects, and for an opinion on Standard oil from Buenos Aires and
the provinces as reported in the Argentine Chamber of Deputies, see memorandum, Antonio Fialho to MRE, La Paz, March 25, 1935, in MRE, *Ofícios La
Paz*, 1935, Arquivo Histórico. For the Standard of Bolivia nationalization
controversy consult Herbert S. Klein, "American Oil Companies in Latin
America: The Bolivian Experience," *Inter-American Economic Affairs* 18,
no. 2 (Autumn 1964): 47–72.

36. Conversation with Souza Costa, May 5, 1941, in Diário, Arquivo Horta
Barbosa.

37. CNP Ofício 9020, Horta Barbosa to Vargas, Oct. 20, 1942 (mimeographed), in Arquivo Horta Barbosa.

38. Dispatch, William C. Burdett to Secretary, Rio de Janeiro, April 9,
1941, DS 832.6363/336; telegram, Jefferson Caffery to Secretary, Sept. 12,
1941, DS 832.6363/396; dispatch 5624, Simmons to Secretary, Oct. 12, 1941,
DS 832.6363/400; Standard Oil *Memorial, Argentine Petroleum Situation*,
Sept. 26, 1940, p. 5, in DS 835.6363/532—all in U.S. National Archives.

39. CNP Relatório 3451, Horta Barbosa to Vargas, July 18, 1941, detailing
his objections to the Standard Oil proposal (mimeographed), Arquivo Horta
Barbosa.

40. Horta Barbosa, Diário for July 25, 1939; on the 29th Mr. Nale of Atlantic Oil came in and spoke about the possibility of financing the refinery
(Arquivo Horta Barbosa).

41. Telegram, Caffery to Secretary, Sept. 12, 1942, reporting conversation
with Aranha, DS 832.6363/462, U.S. National Archives.

42. Martins, *Pouvoir et developpement économique*, pp. 304, 306–307;
Horta Barbosa conversation with Vargas in Diário, May 15, 1941, Arquivo
Horta Barbosa.

43. Robert A. Potash, *The Army and Politics in Argentina, 1945–1962*
(Stanford, 1980), quotation on p. 71; for oil see pp. 67–76.

44. Caffery telegram, cited in note 41, and Horta Barbosa conversation
with Vargas in Petropolis, March 25, 1942, in Diário, Arquivo Horta Barbosa.

45. Max Thornburg letter to Philip Bonsal, from the Office of the Petroleum Advisor, Oct. 28, 1942, DS 823.6363/475, U.S. National Archives.
Thornburg, a former SOCAL executive who later played the key role in
working out the famous 50–50 profit-sharing formula in Venezuela (according to historian Clayton Koppes, personal communication), was no friend of
state enterprise. In fact, major new fields were discovered at the end of the
1930s and YPF brought them into production.

46. Monteiro Lobato considered himself a friend of Vargas and had access

to the president. See letter, Monteiro Lobato to Luis Vergara, São Paulo, April 7, 1938, and letter of Monteiro Lobato to Vargas, March 31, 1938, pleading with him not to sign the new petroleum law. "Essa lei representa logo de entrada o golpe de morte que o Sr. Fleury da Rocha sempre sonhou dar nas companhias nacionais de petróleo. *Esse homem é o agente secreto dos Poderes Ocultos hosteis ao petróleo brasileiro dentro do oficialismo*" (original emphasis). Both letters in Tribunal de Segurança Nacional, Processo 1607, Arquivo Nacional.

47. Relatório, Gen. Francisco Pinto, director of the Conselho de Segurança Nacional (CSN), to Vargas, undated, CSN Arquivo. Vargas approved it on Sept. 28, 1938.

48. For example, the letter of José Faustino to Monteiro Lobato, Pelotas, RS, Feb. 28, 1939. Faustino, who said he served as "your loyal stock agent and was an enthusiast in your great campaign to nationalize petroleum, was inspired by your book and believed in you," could not go along with Monteiro Lobato's directive to sell five hundred shares at a discount to finance a business trip to Argentina. "What will the stockholders think of you—who hold so much stock [50 percent of the Cia. Matogrossense de Petróleo]—based only on rights and contracts with landowners in Mato Grosso, if you then turn around and sell stock?" The CNP's extensive investigation into the tangled affairs of this company is in Ofício 4602, Horta Barbosa to Vargas, Dec. 17, 1940, among many supporting documents in TSN, Processo 1607, Arquivo Nacional. See also the Attorney General Hahnemann Guimarães' official report on Matogrossense, May 19, 1941, reprinted in full in Lourival Coutinho and Joel Silveira, *O petróleo do Brasil: Traição e vitória* (Rio de Janeiro, 1957), pp. 383–94.

49. Letter, Monteiro Lobato to Fernando Costa [intervenor of São Paulo], São Paulo, April 29, 1940, pp. 52–60, in Monteiro Lobato, *Obras completas, cartas escolhidas*, (São Paulo, 1959), vol. 16, book 2. Letter, Monteiro Lobato to Vargas, May 24, 1940, and letter to General Góes Monteiro, May 31, 1940, both in TSN, Processo 1607, Arquivo Nacional.

50. Letter, General Góes Monteiro to Monteiro Lobato, Rio de Janeiro, July 11, 1940, in TSN, Processo 1607, Arquivo Nacional.

51. Letter, Pericles de Magalhães, editor of the *Revista do Petróleo*, to Monteiro Lobato, São Paulo, July 9, 1940, and Monteiro Lobato's reply on the 10th, in the TSN, Processo 1607 documentation. Horta Barbosa on July 26 noted in his diary that the *Revista* was organizing a conference to attack the CNP orientation with respect to national refining and exploration companies. On May 7 he referred to the *Revista's* columns hostile to the CNP, "which doesn't want to give it the money they asked for" (Diário, Arquivo Horta Barbosa).

52. Wirth, *Politics of Brazilian Development*, ch. 8. For a discussion of military politics and the oil issue, see Antonio Carlos Peixoto, "Le Club Militar et les affrontements au sein des forces armées (1945–64)," in *Les*

partis militaires au Bresil, ed. Alain Rouquié (Paris, 1980), pp. 65–104. Martins also speculates on this in his *Pouvoir et développement*.

53. José Murilo de Carvalho, "Armed Forces and Politics in Brazil, 1930–45," *Hispanic American Historical Review* 62, no. 2 (May 1982): 208, 220–21.

54. I am very grateful to Col. Luís Augusto Horta Barbosa for stimulating conversations and many insights into his father's character during October and November in Rio de Janeiro, 1980.

55. Professor José Carlos Sebe bom Meihy, author of a forthcoming book on Monteiro Lobato and a resident of Taubaté, gave me many insights into his illustrious *conterráneo*'s character, thought, and actions.

56. Letter, Monteiro Lobato to Pericles Magalhães, São Paulo, July 10, 1940, in TSN, Processo 1607, Arquivo Nacional.

57. Coutinho and Silveira, *Traição*, pp. 376–78; TSN, decision signed by Colonel Meynard Gomes, April 8, 1941, in Processo 1607, Arquivo Nacional.

58. Daniel Smith, an Americanist at Stanford, provided a searching commentary and critique on these issues (letter, Jan. 31, 1982).

59. Conversation with Vargas, July 9, 1943, in Diário. On the struggle with João Alberto: two memos from Horta Barbosa to Vargas, Oct. 22 and Nov. 6, 1942; a memo from Horta Barbosa to João Alberto, Nov. 26, 1942; and the late November ruling of the National Security Council (which went against Horta Barbosa) in a memo from Gen. Firmo Freire do Nascimento, secretary general of the CSN, to Horta, Nov. 30, 1942. All in Arquivo Horta Barbosa.

60. Conversation with Leslie Webb, July 9, 1943, in Diário. Webb worked for Atlantic after the war. The State Department did not insist on his visit but said that because of demands from other sources the War Production Board might delay shipping the equipment unless it was shown to be really needed and would be put to good use. "The Brazilian reluctance to have another [American] technician visit the area is obviously due to the fact that Kemnitzer [the Cooke Mission's petroleum specialist] had very optimistic ideas as to the future of the Bahia development which have not been borne out in practice. They fear that another investigation might result in their not getting the equipment." Despite this sceptical (and biased) assessment, the United States had no intention of canceling the $600,000 worth of equipment, in addition to $200,000 worth already released from government funds for the project (memo, Allan Dawson to Philip Bonsal, July 22, 1943, DS 832.6363/623, U.S. National Archives). For the Cooke mission consult Frank D. McCann, *The Brazilian-American Alliance, 1937–1945* (Princeton, 1973), pp. 382–88.

61. Maurice Pearton, *Oil and the Romanian State* (Oxford, 1971), pp. 237–39.

Esperanza Durán

Pemex: The Trajectory of a National Oil Policy

Introduction: Mexico—Oil Rich, Again

The discovery of vast oil deposits in the mid-1970s made Mexico a major oil power overnight and was greeted as a remedy not only to a severe national economic crisis but also, especially, to an ailing Pemex, the state oil monopoly. By 1980, with proven reserves of fifty billion barrels and potential reserves of two hundred billion barrels, Mexico's role as a major non-OPEC supplier to the United States, Japan, and Western Europe was substantial.[1] Increased stature in the international petroleum market brought about a major shift in oil policy. Moving rapidly away from its postexpropriation goal of autarchy, centered on energy self-sufficiency and economic independence, Mexico placed the weight of its development strategy on oil, which became its chief export and main revenue source (tables 1 and 2).

This rather sudden shift in emphasis away from internal concerns to foreign markets was a direct consequence of the massive oil discoveries. But it also coincided with changes in the Mexican economy, which faced a payments crisis. Old strategies based on import substitution and growth behind tariff walls were no longer viable. The nation was rethinking its development strategy and thus the extent of its interaction with other economies.[2] Having a large industrial base already in place meant that the country could absorb a large inflow of petrodollars, and no particular danger in expanding oil exports was perceived.[3] Unfortunately, the danger of basing the new growth and development strategy on petroleum exports was all too clear by mid-1981, when the sudden fall in oil prices staggered an economy that depended on Pemex oil for 75 percent of its exports.

4. Mexican workers on a union march through the streets
of Tampico, about 1921. Courtesy M. Philo Maier.

This was not the first time that oil exports had thrived. Mexico was a major oil exporter for the first two decades after its revolution, mostly to the United States, which, while a net exporter itself, did not produce enough heavy oils for its own consumption. Between 1913 and 1918 over 95 percent of United States imports of heavy petroleum came from Mexico.[4] Foreign capital controlled the industry in uneasy relationship with the revolutionary leader. Porfirio Díaz had encouraged foreign investment in the oil industry by offering political stability, generous concessions, tax exemptions, and subsidies. To these laissez-faire Díaz policies the new revolutionary leadership reacted aggressively, along nationalist lines. Oil became a symbol of national independence, as well as a means to pay for social benefits. The control of natural resources touched one of the most sensitive chords of the young revolution. But if the ideal was nationalism and social justice, the reality was that this large industry remained in foreign hands.

Circumstances led to the final collision of private interests and the state in 1938. On the one hand, the revolutionary government was stronger and more stable. On the other, the foreign oil concerns were, in the Mexican view, blocking social reforms while continuing to concentrate on exports for profit rather than for the benefit of local industry. Furthermore, they had let the industry run down while turning to more profitable investments in Venezuela and the Middle East. Mexico under Lázaro Cárdenas believed it had no option short of expropriation to control this strategic sector and to assert policies of economic nationalism and independence.

Henceforth, the oil industry was seen as an integral part of the nation; its role was not to make profits in and of themselves but to serve society as a whole. Between 1938 and 1976, the prime objective was to satisfy internal demand and to support import-substitution industrialization through very low, subsidized prices. Oil was regarded as a tool for inward-looking development, a means to supply industry and consumers with cheap fuel, but hardly as an asset in its own right. Why was oil not used as an export to help finance development, in addition to these internal uses? Perhaps it was through ideology or even prejudice, rather than from rational economic decision making. Since oil was at the inner core of nationalism, to export would be to sell off parts of the patrimony. As for importing oil—as happened in the early 1970s—this was tanta-

Table 1: Evolution and Participation of Petroleum Products
in Mexico's Total Exports (millions of pesos)

	Total Exports		
	Value	Annual variation	
		(a)	(b)
1934	643.7	—	
1935	750.3	16.6	
1936	775.3	3.33	
1937	892.4	15.1	
1938	838.1	−6.1	
1948	2,661.3		12.2
1958	8,846.1		12.8
1968	14,758.9		5.3
1969	15,111.8	2.4	
1970	17,162.0	13.6	
1971	18,430.8	7.4	
1972	20,926.5	13.5	
1973	25,880.8	23.7	
1974	35,624.6	37.6	
1975	35,762.9	0.4	
1976	51,905.4	45.1	
1977	94,452.5	82.0	
1978	128,853.9	36.4	
1979	196,434.2	52.4	
1980	397,994.5	102.5	
1981	—	—	

(a) Yearly variation.

(b) Yearly variation average over previous ten-year period.

Sources: Banco Nacional de Comercio Exterior, *México exportador*
(Mexico City, 1939), p. 39; Nacional Financiera, *50 años de Revolución
Mexicana en cifras* (Mexico City, 1963), pp. 139–40; Mexico, Secretaría de

	Oil Industry		
Value	Annual variation		
	(a)	(b)	% of total exports
149	—		23.1
145	2.7		19.3
146	0.7		18.8
151	3.4		16.9
112.0	−25.8		13.4
339.0		13.5	15.0
320.0		−2.2	3.6
538.0		5.3	3.6
535.0	−0.6		3.5
504.0	−5.8		2.9
433.0	−14.1		2.4
324.0	−25.2		1.6
442.9	36.7		1.7
1,668.2	276.6		4.7
5,288.0	217.0		14.8
7,003.0	32.4		13.5
23,431.0	234.6		24.8
41,795.8	78.4		32.4
91,690.9	119.4		46.7
239,502.8	161.2		60.2
357,537.8	49.3		—

Programación y Presupuesto, La industria petrolera en México (Mexico City, 1980), p. 25; Petróleos Mexicanos, Memoria de labores 1981 (Mexico City, n.d.), p. 124. Values have been reproduced in current pesos as given in sources. Conversions to real or dollar values can be performed by the interested reader.

Table 2: Participation of Pemex in the Federal Public Sector's Gross Revenues (millions of pesos)

	Consolidated Public Sector	Nationalized Industries	Pemex	Percentage
1974	186,780.8	115,431.3	32,004.8	10.6
1975	249,425.8	156,304.0	39,774.0	9.8
1976	310,605.3	176,684.3	48,118.4	9.9
1977	437,932.9	247,860.7	70,118.0	10.2
1978	590,403.4	334,111.0	106,097.0	11.5
1979[a]	787,454.0	460,824.0	177,322.0	14.2
Annual Variation (%)				
1974	33.5	33.4	24.3	
1975	33.5	33.4	41.0	
1976	24.5	13.0	21.0	
1977	41.0	40.3	45.7	
1978	34.8	34.8	51.3	
1979	33.4	37.9	67.1	

[a]Estimated figures.

Source: Mexico, Secretaría de Programación y Presupuesto, Industria petrolera, p. 50.

mount to a loss of sovereignty. Because self-sufficiency was the rule, exports and imports were both minimized.[5]

One notes the sharp contrast between this autarchic period and the two export phases, which differ not in foreign trade policy as such but in ownership and control. Whatever the period, be it autarchy or petroleum-led growth by exports as in the recent oil boom, the symbolic place of oil in Mexican nationalism has remained a constant. A vivid current example is provided by Jorge Díaz Serrano, the ex-head of Pemex and the first major victim of President Miguel de la Madrid's strong campaign against corruption. Díaz Serrano is not a member of the old political class, but rather the keeper and by example the betrayer of a most potent symbol of Mexican independence and progress.

Another historical feature of the state oil industry is that no major figure emerged in the early years to lead it, no technocrat as hero. The three other countries covered in this book each produced a dominant figure who greatly influenced the national approach to oil, the structuring of the industry, and its relation to the rest of the economy, including domestic private capital. Mexico had no Enrique Mosconi, Julio Caetano Horta Barbosa, or Juan Pablo Pérez Alfonzo. Nor was one needed, for it seems that oil in Mexico became a national project, with its own mystique that supports consensus. This national project was forged during the twenty years before nationalization, in the almost continuous conflict of interest between revolutionary nationalism and the foreign oil concerns. When Cárdenas expropriated the industry in 1938, he also incorporated the nationalist ideals and experiences of his predecessors.

To be sure, the oil nationalists did not have a long-term plan on ways to achieve their objectives. They had no clear ideas on such important issues as the level of oil production, how to balance internal consumption with exports, or the role of private capital in the industry. As with most policies of this regime, the role of trial and error and changing priorities yielded different decisions in different periods.

Genesis of Oil Nationalism

The early petroleum industry in Mexico was pioneered by foreigners. Before 1900 the most important concern was the Waters-Pierce Oil Company, directed by Henry Clay Pierce, an American whose other Mexican interests included mining and railroads. Standard Oil owned 65 percent of the company, Pierce the rest. Waters-Pierce had a monopoly over the sale and distribution of United States refined products in Mexico; it did not produce oil. Taking full advantage of his market position, Pierce charged very high prices. Standard disapproved, expecially since the state of Texas was investigating him for profiteering and unfair market practices. In 1900, Waters-Pierce was expelled from Texas. Shortly after the Standard Oil Company was broken up under antitrust legislation in 1911, Pierce bought out Standard and became full owner.[6]

Actual oil production was started by Edward L. Doheny, a California oilman who acquired 280,000 acres in the Tampico area and began drilling in May 1901. By 1904 his Mexican Petroleum Company (organized under the laws of California in 1901) and its subsidiaries, the Huasteca Petroleum Company and the Tamiahua and Tuxpan companies, were drilling with success along the so-called *Faja de Oro* or Golden Lane, a coastal strip from northern Veracruz to Tamaulipas. Doheny was the first to produce on a commercial scale.[7]

American dominance of the fledgling Mexican oil industry was challenged by Sir Weetman Pearson (created Lord Cowdray in 1911), a British subject who had been encouraged by President Díaz to ease dependence on American investment. Indeed, despite Díaz's popular image as the man who opened Mexico to American investment and trade, he was conscious of the dangers of relying too heavily on the United States as Mexico's senior partner and thus sought alternative capital. Furthermore, Díaz esteemed Cowdray as a master builder, the constructor of Dover Harbor, the Blackwall Tunnel under the Thames, and New York's East River Tunnel; closer to home, he had developed the Mexico City sewerage system and the port of Veracruz and had rebuilt the Interoceanic Railway of Tehuantepec.[8]

To fuel this railway, Cowdray attempted to develop oil seepages in the Tehuantepec area and built the first refinery in Mexico at Minatitlán in the hope of marketing petroleum products on a large scale. But oil production fell short of expectation. Under contract to supply a British distributor in Europe, he was forced to buy Texas oil, refine it in Mexico, and reship it to Europe. An optimist when it came to Mexican oil potential, Cowdray then brought in several

good wells on his lands along the Golden Lane. Late in 1910 his
Eagle Oil Company (El Aguila, or Compañía Mexicana de Petróleo)
was producing 60 percent of Mexican oil.[9] Believing that the company should be owned by Mexicans, Cowdray registered El Aguila
in Mexico and put several leading politicians, including Díaz' son,
on his board. Despite his expectations, the first public stock issue
did not sell well in Mexico, and most of the shares were sold in
Europe. The lack of interest on the part of Mexican capitalists was a
harbinger of things to come, for by 1938 independent Mexican producers in the Tampico area accounted for less than 5 percent of total
oil production.[10]

The success of Doheny and Cowdray soon attracted the major
American oil companies, which established Mexican subsidiaries:
Mexican Gulf Oil Company, the Texas Company, the Tampico Fuel
Oil Corporation (Sinclair), the Panuco-Boston Oil Company (Atlantic Refining), and Transcontinental (Standard Oil). La Corona was
owned by Royal Dutch–Shell. Mexico became the world's largest
oil exporter, and even the outbreak of revolution did not cause oil
investment to slacken. Mining investment did go down, but the demand for oil rose during World War I and this encouraged oilmen to
keep investing in Mexican oil.

Destruction of the oil fields and installations during the revolution was minimal. Manuel Peláez, a local caudillo, secured them.
Being located in an enclave, the oil properties were easy to guard,
whereas the spread-out mining installations did not fare so well.
In the early revolutionary phase (1912), mining investments amounted to about $250 million, with oil far behind at $15 million. But by
1919 oil was up to $200 million, whereas mining totaled only $222
million.[11] Oil profits soared; by 1919 they reached 13 million pesos,
or 10 percent of national income, exceeding mining revenues
by one million pesos.[12] This figure was not much higher in 1938.
(In the recent export boom, Pemex' share in the government's gross
revenues increased 67 percent between 1978 and 1979, when
they accounted for 14.2 percent of the public sector's total gross
revenues.)

Before the 1938 expropriation there were two main sources of
conflict between the oil companies and the government: nationalist
clauses in the 1917 Constitution and tax policies of the revolutionary governments. That the revolution had ushered in a new era for

the companies was manifested first in the tax realm, although the new constitution set in motion legal changes that in the long run were even less to the companies' liking.

Oil production began under terms of the liberal Mining Code of 1884, which was based on the Anglo-Saxon legal concept of absolute ownership of the subsoil. The foreign companies bought or leased large tracts on the assumption that the mining code gave them the right to exploit in perpetuity whatever petroleum they found. The industry grew rapidly, and nationalist revolutionaries began to discuss how petroleum could revert to federal ownership. Various attempts to formulate a new policy culminated in the famous Article 27 of the 1917 Constitution, which reasserted the old Hispanic principle of sovereignty over subsoil deposits.[13] When this principle was applied retroactively, the companies claimed prior rights and said this contradicted Article 14 of the same constitution banning retroactive legislation. President Venustiano Carranza chose to interpret the law in a somewhat different manner: under Article 27 the subsoil was not reclaimed by act of expropriation because the nation had *always* owned it, as the preamble of Article 27 stated.[14]

In the absence of enabling legislation, the new constitution had no immediate effect on property rights. By the time this was enacted in 1925 the companies had been protesting repeatedly to the government and their embassies against the constitution, as well as against new tax policies that were having a much more immediate effect.

Taxes on the oil industry were virtually nonexistent before the revolution. To stimulate development, Díaz allowed all materials and equipment to enter duty free while charging producers an insignificant stamp tax. Under Francisco I Madero, the first revolutionary president, oil became a source of revenue to the government for the first time, the tax being 20 centavos per ton of crude (see tables 3 and 4).[16] In 1912, the price of oil was three pesos in the well, and the total tax amounted to 28 percent of the producer's price.[17] During Victoriano Huerta's administration the federal tax was increased to 75 centavos. Now the total tax came to half of the gross price of oil in the well, and the oil companies were finding conditions less favorable. The Constitutionalist faction, which took power after Huerta's fall, reestablished the 20 centavo tax but levied it in gold to compensate for a rapidly depreciating currency. The

Table 3: Tax Revenues from Production of Oil
and Derivatives before Expropriation (pesos)

	Federal	State	Municipal	Total
1912	494,275			494,275
1913	767,043			767,043
1914	1,232,930			1,232,930
1915	1,942,687			1,942,687
1916	3,088,368			3,088,368
1917	7,074,968			7,074,968
1918	11,480,964			11,480,964
1919	16,690,622			16,690,622
1920	45,479,168			45,479,168
1921	50,604,049			50,604,049
1922	58,177,029	197,127		58,374,156
1923	40,042,073	2,100,649		42,152,722
1924	37,071,041	1,881,695		38,952,736
1925	28,846,380	1,519,684		30,366,064
1926	22,228,170	2,453,398	15,904	24,697,472
1927	11,941,574	1,190,969	132,330	13,264,873
1928	6,438,204	643,821	71,536	7,153,561
1929	4,653,022	465,302	51,700	5,170,024
1930	4,614,943	461,323	51,258	5,127,524
1931	3,422,232	342,013	38,001	3,802,246
1932	4,074,808	407,208	45,245	4,527,261
1933	4,467,654	446,414	49,596	4,963,664
1934	7,245,856	724,154	80,462	8,050,472
1935	5,975,430	597,451	66,383	6,639,264
1936	7,633,606	673,361	84,818	8,481,785
1937	9,532,059	953,206	105,912	10,591,177

Source: Mexico, Secretaría de Hacienda y Crédito Público, Departamento
de Impuestos Especiales, in Miguel Manterola, La industria del petróleo
en México (Mexico City, 1938), p. 387.

State of Veracruz imposed its own tax on production, and when
Carranza had to flee Mexico City temporarily he established "bar
duties" on all petroleum exports from Tampico.[19] The Constitu-
tionalists pushed up oil taxes from then on. As one foreign observer
put it, "Under Carranza the principle of taxation was that of the

Table 4: Participation of Taxes on Petroleum and Its Products in the Total Tax Revenue of the Federal Government[a] (millions of pesos)

	Total Tax Revenue of Federal Government	Specific Taxes on Petroleum and Its Products			Taxes Paid by Pemex[b]		
		Value	Annual Variation[c]	Participation in total (%)	Value	Annual Variation[c]	Participation in total (%)
1938	362.8	18.4	—	5.1	18.4	—	5.1
1948	1,874.2	143.4	22.8	7.7	143.4	—	7.7
1958	8,502.2	522.6	13.8	6.1	522.6	13.8	6.1
1968	28,365.3	1,206.3	8.7	4.3	1,166.2	8.7	4.1
1969	32,033.5	1,409.3	16.8	4.4	1,368.3	17.3	4.3
1970	36,624.4	1,613.3	14.5	4.4	1,571.9	14.9	4.3
1971	40,057.4	1,245.3	(22.8)	3.1	1,204.5	(23.4)	3.0
1972[d]	47,445.8	3,795.8	204.8	8.0	3,749.0	211.2	7.9
1973	62,494.7	2,127.3	(44.0)	3.4	2,073.0	(44.7)	3.3
1974	91,238.7	5,466.0	156.9	6.0	4,897.6	136.3	5.4
1975[d]	124,701.4	14,308.3	161.8	11.5	8,598.3	75.6	6.9
1976	154,796.7	14,454.7	1.0	9.3	7,760.9	(9.7)	5.0
1977	218,383.0	28,905.1	100.0	13.2	18,252.7	135.2	8.3
1978	294,817.3	40,457.4	40.0	13.7	28,299.7	55.0	9.6
1979	397,023.0	62,969.4	55.6	15.9	48,385.0	71.0	12.2

[a]Gross Figures.

[b]Excluding specific taxes. Consists primarily of extraction, production, and export taxes.

[c]For 1948, 1958, and 1968 variations mean average rates of annual growth during the previous ten years.

[d]In 1972 and 1975 the significant growth of taxes on oil and its products is due to complementary payments made by Pemex worth 1,399.2 million pesos and 1,066.0 million pesos respectively that were compensated by federal government subsidies.

Source: Mexico, Secretaría de Programación y Presupuesto *La Industria petrolera*, p. 51.

Irishman in the fight at Donnybrook Fair: 'Whenever you see a head, hit it.' "[20]

In 1915, Carranza created the Petroleum Technical Commission (CTP) to implement government policy. Cándido Aguilar, the Vera Cruz governor and Carranza's son-in-law, presided over it along with Pastor Rouaix, a staunch *carrancista*, leader in the Constitutional debates, and minister of industry, and Modesto Rolland, Miguel Urquidi and Salvador Gómez. The engineers Joaquín Santaella and Alberto Langarica were technical assistants on the CTP. All commission members were young professionals who sympathized with the revolutionary ideology. With their middle class orientation they typified the social background of the Constitutionalist forces. As the historian Friedrich Katz has noted, "the heavy participation of members of the middle class in the civilian and military leadership of the Carranza movement is evidence of the extensive influence they wielded in that movement, not least because the revolution led by Carranza offered the Mexican middle class access to the elite political, military, and financial positions of the country. His nationalism particularly reflected middle class fears of growing foreign domination."[22] Significantly enough, it was just this class that provided the backbone of support for Enrique Mosconi's YPF (Yacimientos Petrolíferos Fiscales) in Argentina.

These middle class intellectuals and professionals gave the revolution its ideological content, with a heavy dose of nationalism and the idea that the state should have greater control of the economy. The 1917 Constitution in general and the oil legislation in particular reflected the thinking of an incipient technocracy whose members occupied high posts in the revolutionary governments of that period.

The CTP mission was to evaluate the industry and all matters relating to it, and then to draft new laws and regulations. It began by revising oil taxes and by decreeing new rules for concessions and oil production.[23] These measures the foreign oil companies deeply resented, and they protested constantly to their foreign offices against arbitrary, unjust, and confiscatory acts by the Mexican authorities. The decree of January 7, 1916, was particularly galling: this ordered the suspension of all industrial activity—from production and exportation to the drilling of new wells and the laying of pipelines—until new legislation was drawn up specifying the legal status and fiscal regime under which the industry would operate.

Oilmen objected immediately, and Lord Cowdray told the British authorities that because of this decree El Aguila was losing up to 100,000 pounds sterling a day.[24]

Acts by the government that the oil companies considered hostile continued all through 1916 and 1917. The contracts and concessions of some companies were canceled.[25] Exploration and production were suspended several times despite company protest. On top of all this, it was decreed in early 1917 that any foreigner who acquired real estate or concessions to exploit Mexican natural resources would have to forfeit the protection of his government and consider himself juridically a Mexican citizen.[26]

As conflict mounted, the decree most opposed by the companies was issued on February 19, 1918, to enforce Article 27 of the constitution. This decree also imposed new taxes on oil production and refining as well as on the ownership or lease of oil properties, without exempting rights acquired before the constitution. But the most significant part of this decree, and the one the companies feared the most, was Article 14 stipulating that the companies had to register their property titles or lease contracts in the Department of Industry, Commerce, and Labor. In case of noncompliance, the oil properties would be considered vacant and open to "denunciation" by third parties.[27] To the operators, this seemed a device to get them to endorse implicitly Article 27 of the constitution. Attorneys for the companies were convinced that Article 14 of this decree clearly indicated the intention of Mexican authorities to separate ownership of surface property from the subsoil mineral deposits. Registration would therefore imply legal acknowledgement of this principle.[28] Feeling threatened by this decree, the companies closed ranks. El Aguila asked Chandler P. Anderson, attorney for the United States companies, to represent it as well.[29]

Yet the British and American companies had different views on how to protect their interests. Lord Cowdray thought that the best way to guarantee his Mexican properties was for his government to grant complete diplomatic recognition to the Carranza government, as the United States government had done formally in August 1917. American oilmen preferred the removal of Carranza. They pressured the State Department to take effective measures—including threats of intervention—to force Carranza to adopt a more benevolent attitude toward American interests. They insisted that complete recognition be withheld until Carranza promised not to make Article 27 retroactive.[30]

Obviously, Mexico had the oil companies under severe pressure. Despite this, conventional scholarship on this period has assumed that the government was not powerful enough to control the companies until expropriation, in 1938. A different interpretation would seem appropriate. We have seen that Carranza's government was indeed willing and able to impose substantial new directives on the industry. Note also Jonathan Brown's chapter in this volume, which shows that Standard Oil was unable to counter assertive actions by national governments. Thus strong action by the state was possible. The hard part was first to design a policy and then to follow through. In essence, the real problem facing Mexico was not the power of the companies but how to develop an effective state oil policy.

According to George Philip, oil nationalism in Latin America "can be understood only if we bear in mind the fact that [oil nationalist movements] represent in large part an assertion of the rights of the central government over those of provincial authorities or landowners."[31] Carl E. Solberg and John D. Wirth provide data in their chapters to support this. Attacks on the central government, perceived as affronts to the country, became tools of nationalism in the hands of leaders seeking to unite and control the country. In Mexico, memories of the risks and dangers stemming from a weak central government, as in most of the nineteenth century, were still fresh. Oil became a symbol and a vehicle for national unity. The fires of petroleum nationalism were fanned by the actions of Peláez, the Vera Cruz caudillo who successfully defied Carranza. This episode left a profound imprint on the revolutionary leadership as well.

As military protector of the foreign oil companies during the later years of armed revolution, years which coincided with the great spurt of demand for Mexican oil generated by World War I, Peláez really shook the revolutionary elite. The importance of this episode was related by Cárdenas twenty years later in his expropriation message. "The oil companies' support for strong rebel factions against the constituted government in the Huasteca region of Veracruz . . . during the years 1917 to 1920 is no longer a matter of dispute by anyone."[32] The Peláez case reverberated throughout Latin America, being held up by nationalists as a classic example of danger to the state when foreign companies collude with a private regional army.

Just how this famous example of foreign support for regional autonomy came about is a matter of some interest. The power of Peláez derived from his ability to occupy oil fields where the interests of several major actors coincided. These included the British and American oil companies, the British Admiralty which relied heavily on Mexican oil supplies, and the British and American foreign offices charged with defending both private and national interests.

But the triangular relationship established between Peláez, the oil companies, and the already legitimized revolutionary government of Carranza, is indeed a complicated one. On the one hand, the companies were at Peláez' mercy, since his military occupation of the oil fields allowed him to disrupt or obstruct production at · any time or even to sabotage vulnerable installations. Thus it was in their interest to keep Peláez happy. But they also used him as a lever against Carranza, who could not secure the oil-producing area. And for Carranza oil had become an important revenue source (see table 3) and it was in *his* interest to keep the oil flowing; occupation by Peláez was in fact preferable to a halt in production. As for Peláez, cash from the companies allowed him not only to remain independent of the central government but also to help other regional caudillos who wanted to overthrow Carranza, such as Emiliano Zapata, Félix Díaz, Guillermo Meixueiro, and the Cedillo brothers.[33] As he became established as a prosperous revolutionary, Peláez used his economic power to pay soldiers heretofore loyal to the other chieftains who found it difficult to feed and arm their men.[34]

Before becoming involved in these military activities, Peláez was a landowner in the Huasteca region (parts of the states of Tamaulipas, San Luis Potosí, Vera Cruz, and Hidalgo), and some of the principal oil wells in the region were located on lands leased from his family. Peláez had once supported the carrancista faction, but in 1914 he opted for Pancho Villa and the Convention.[35] In November of that year Peláez and his followers based themselves in the oil-producing area and began to levy forced "loans," which became a regular source of income and which the companies paid in return for protection. Lord Cowdray was prepared to pay blackmail "in moderation" to protect his wells from stoppages that would ruin them as well as threaten an entire oil field. The State Department also reckoned that the best policy was to pay, though under protest.[36]

The paying of money and, later, of arms and ammunition to Peláez was something the British and American authorities were aware of, and also encouraged. Referring to the United States, the historian Robert Freeman Smith rightly points out that "the State Department . . . did not object—it even gave quiet support—to the private 'diplomacy' of the oil companies." This was awkward in view of the fact Peláez was in open revolt against a central government recognized by the United States. Just how much money he received is hard to estimate.[37] But that the arms and ammunition he received directly from the companies enabled him to control the area is clear enough, although direct evidence of his receipt of arms is lacking. In February 1917 Harold Walker reported to the State Department that the Huasteca Petroleum Company he represented was under heavy pressure to ship rifles and cartridges to Peláez. Chandler Anderson, lawyer for American and British interests, wrote in his diary for March that "[Secretary of State Robert] Lansing had told him that the U.S. government would not prosecute anyone who supplied arms and ammunition to Peláez."[38]

As unease grew that German forces might raid the oil fields, it seemed even more natural to provide Peláez, who always wanted more arms, with "the sinews of war." The American companies, Mexican Petroleum especially, saw aiding Peláez as the best way to protect their oil fields. The State Department also considered him "the man most likely and best able" to afford protection and informed the British Embassy in Washington that when it came to sending arms to Peláez their eyes were shut. Of course the department wanted this held in secret, for it ran the obvious risk of being charged with actions against the government it recognized.[39] Lord Cowdray sent money but stopped short of supplying arms, recommending to the Foreign Office instead that it would be easier for Americans to furnish the munitions.[40]

Carranza's forces skirmished with Peláez but could not dislodge him, and by late 1917 the president's keen desire to do so was being reported openly in the British and American press.[41] Apart from the common "German intrigue" theory, it was believed that Carranza wanted to control the oil fields in order to threaten massive new taxes as a way to force Washington to lift the embargo on the export of United States foodstuffs and gold to Mexico.[43] The upshot was that nothing changed until Carranza was assassinated in May 1920, at which point Peláez promptly ended his revolt and embraced the Agua Prieta Plan being pushed by the so-called Sonoran dynasty of northern generals.[43] These generals under Alvaro Obregón and Plutarco Elías Calles led what amounted to yet another cleavage in the revolutionary movement, accusing the central government of making popular suffrage a farce, of violating state sovereignty, and of betraying revolutionary principles that were "already raised to near sacred status."[44] Within weeks the Sonoran dynasty controlled the country. The whereabouts of Peláez after this are not well established, but further research on this important revolutionary caudillo would quite likely raise more questions about the role of "independent" leaders in a revolutionary situation.

The demand for gasoline and fuel oil grew rapidly after World War I, propelled by the auto industry and the conversion of ships from coal to oil. The inability of United States oil producers to meet demand opened space for Mexican imports. But steadily rising prices, and the effects of the Internal Revenue Act of 1918 (which allowed the "discovery value" instead of cost value as a basis for depletion) spurred exploration both at home and abroad. The result was a series of United States oil discoveries that ushered in a period of chronic oversupply, starting in late 1920 and continuing almost uninterrupted until 1935.[45] This helps explain why after peaking in 1921 Mexican production steadily declined. As for global market factors, it is well worth recalling that in 1928 the major oil companies worked out a scheme to avoid ruinous competition in a situation of oversupply. This was the Achnacarry Agreement, based on the "as is" principle under which the majors agreed to maintain a constant market share in each country where they operated. Perhaps even more important, Achnacarry provided for a common pricing system and a framework for specific cartel arrangements.[46]

In the 1920–28 period, four events had a decisive impact on the company-state relations in Mexico. The first was the "positive act" decision of 1922; in this the Mexican Supreme Court decided in favor of the Texas Company, which had challenged the retroactivity of the 1917 Constitution. The court specified conditions whereby a prior "positive act," or evidence that the companies really did intend to produce oil on properties they held, could create vested interests. The second policy determinant stemmed from the Bucareli Conference between Mexico and the United States over the recognition of Obregón's government. Then came the enactment of the Petroleum Law of 1925, the first bylaw to implement definitively Article 27. Finally, an agreement between United States Ambassador Dwight Morrow and President Calles modified the 1925 law: concessions without time limits would be confirmed only if a company could demonstrate the performance of positive acts before May 1917.

The fall from power and death of Carranza in May 1920 and the ascent of the Sonoran dynasty ushered in a more cooperative phase in company-state relations. All the pending oil company claims and the cluster of regulatory decrees were inherited by the interim Adolfo de la Huerta government, and later on by the newly elected government of Obregón. Obtaining United States diplomatic recognition in order to stabilize his government was Obregón's immediate problem. American recognition was a sine qua non for Mexico to regain its place as a viable nation in the international community,[47] since the European governments were following the United States lead in formulating their policies toward Mexico. Obregón knew that recognition would not be forthcoming as long as American investors charged the Mexican government with following policies designed to confiscate foreign property. Washington welcomed Mexico City's statement that the government had not confiscated and did not plan to confiscate any legitimately acquired property and that, furthermore, Mexico would honor all international obligations.

Powerful interests, however, tried to prevent premature recognition and demanded first an explicit statement from the Mexican authorities that American rights and properties would be respected. In the course of its investigations into the damages American citizens had suffered during the Mexican revolution, the Fall Committee of the United States Senate advised the new Warren Harding administration not to grant recognition until the Obregón government agreed to exempt American citizens from the effects of certain articles of the Mexican Constitution, namely Articles 3, 27, 33, and 130.[48] The committee felt that recognition should follow a treaty of amity and commerce providing specifically that property should not be subject to confiscation; that the right of expropriation would only be resorted to on grounds of public utility and should include prompt payment and just compensation; and finally that the Mexican Constitution not be applied retroactively.[49]

Obregón refused to sign this treaty as a precondition for recognition, but he did take a conciliatory stand toward the American investors and their government. While reaffirming that "the natural resources of the nation belong to the nation," Obregón also gave assurances that "the famous Article 27, one clause of which declares national ownership of subsoil rights, in petroleum will never be given retroactive effect."[50]

During Obregón's first two years in office drilling permits again were issued, thus lifting the ban Carranza had imposed when the companies refused to comply with his decrees. He made this conciliatory gesture to revive exploration. However, Obregón issued these permits on a temporary basis until the Mexican Congress enacted the legislation necessary to implement Article 27.[51] He also decreed a new export tax to take effect on July 1, 1924, including new rates per cubic meter on all petroleum—crude, fuel oil, diesel, gasoline, kerosene, and lubricants. Although no corroborative evidence has been found, it has been argued that the new export tax did not annul existing oil taxes and that this was Obregón's way of pressuring American investors, thus acquiring more bargaining power to reach a favorable settlement of Mexico's foreign debt.[52] The official justification, as expressed by the chief of the Petroleum Technical Commission, was that the new tax would be used to achieve three objectives: to help repay the national debt, to discourage overproduction, and to prevent the United States from establishing an import tax on Mexican oil imports in competition with American petroleum.[53]

As might be expected, the new tax was protested by the oil companies, which decided to send to Mexico City a delegation from the Association of Producers of Petroleum in Mexico. The importance of Mexican oil to the companies was shown in the high level of this delegation: Walter Teagle, president of Standard Oil of New Jersey; Edward L. Doheny, president of the Mexican Petroleum Company; J. W. Van Dyke, president of Atlantic Refining; Harry Sinclair, chairman of Sinclair Consolidate; and Amos Beatty, president of the Texas Company.[54] The oilmen conferred with Secretary of the Treasury de la Huerta in late August and reached agreement on September 3, 1921. The terms were not made public, except that tax payments would be made quarterly instead of monthly. To forestall domestic criticism, the government announced (inaccurately) that oil taxes would not be reduced, although in fact it had found a way to do just that, at a profit. Early the following year, a presidential decree announced that export taxes could be paid either in gold or in bonds of the public debt, bonds that would be honored at their nominal value, which in effect amounted to a 40 percent tax reduction.

But this compromise between oilmen and the Mexican state was not to become reality. Thomas Lamont of the Bankers' Association intervened and convinced Secretary of State Charles Evans Hughes that the Mexican tax agreement was detrimental to American bondholders and bankers. The oil companies were forced to give up their benefits and pay the higher oil taxes. Thereafter, the higher petroleum revenues helped pay off the foreign banks that held Mexico's defaulted debts.[55]

Company representatives met again the following year with Mexican officials in Mexico City and New York not only to settle the tax question definitively but also to discuss in general the state of the petroleum industry in Mexico.[56] Although the tax question was resolved, the American government still did not recognize Obregón. Then, in the summer of 1922 (following a bankers' visit to Mexico City in November 1921), an accord was reached in New York to refund the Mexican debt in what became known as the de la Huerta-Lamont Agreement. The debt settlement cleared the way for the Bucareli Conference of 1923, leading at last to American recognition.

As a result of the Bucareli Conference, both governments agreed to compromise: no treaty of amity was signed (as the Fall Committee had proposed), but neither did Obregón receive unconditional recognition. On oil matters, the final agreement reflected give and take on both sides. American negotiators agreed that titles to oil properties could be transformed to "confirmatory concessions," which in effect meant accepting a change in their legal status. The Mexicans, in turn, broadened the definition of a "positive act" so that almost all the oil properties and leases would be covered by it.[57]

Although the Bucareli Agreement eased tensions between the two governments, the oilmen were far from satisfied. Mexico had acquiesced in the transformation of their property titles into concessions; the companies still maintained that they *owned* the oil deposits.[58] New complications arose shortly after Obregón transferred the presidential sash to Calles, who promptly revived the oil conflict by drawing up legislation to implement Article 27. This bylaw, enacted on December 27, 1925, affirmed the principle of national direct ownership of the oil deposits as "inalienable and imprescriptible." The oil industry was declared a public utility, and oil concessions were henceforth to be granted separately for exploration and for production.

Concerning the status of oil fields predating the constitution, this new law stated that all operations or contracts signed for the express purpose of developing oil (i.e. the "positive act") before 1917 were valid, but not for more than fifty years from the date production was started or the contracts signed. Moreover, the oil companies would have to apply for confirmation of these rights within a year; failure to file this claim would be assumed by the government as a renouncement of these rights.

Pressures from the oil companies, especially the Americans, through their governments brought about a major change in this law. The appointment of Morrow as United States ambassador, and the cordial relations he was able to establish with Calles, paved the way. Following Morrow's advice, Calles tried to assuage the companies and their hostile attitude toward the Mexican government by revising the 1925 law. Based on the precedent established in 1922 by the Texas Oil Company Case, the Supreme Court ruled that limiting the concessions to fifty years would give this law a retroactive effect. The court also declared unconstitutional certain articles of the 1925 law. Then Congress, acting on Calles' request, revised the law. Finally, by a decree of January 3, 1928, oil concessions were confirmed indefinitely, and the companies had an extra year to confirm their rights.

From the Mexican perspective, these changes upheld the rights of foreign companies to the richest oil territories then known. Most of the properties and concessions on the Golden Lane were obtained before May 1, 1917. According to one interpretation, this decree rendered "theoretical the revindications sought by the constitution."[59] Yet this overlooks the fact that the older wells in the Golden Lane were giving out or were already pumping salt water. After 1921 production declined steadily. (Thirty years later, in the post-Bermúdez era when the nationalized oil industry became a viable enterprise, Mexican production still did not exceed one-half the 1921 level.) To be sure, this decline resulted also from oversupply in the world market, increased by the new discoveries in Venezuela.[60]

In modifying the 1925 law by the decree of January 1928, Calles aimed to create incentives for the companies to increase exploration and output, so the government could raise more revenue. When production did not increase, the authorities blamed the companies for a lack of interest in producing Mexican oil. It is true that the companies were not finding Mexico the most ideal place to do business, because of their uncertain status and the fear of expropriation, but it is also true that they still wanted to tap new petroleum wealth in Mexico. As J. Richard Powell explains, "exploratory activity apparently did not decrease following the peak year of production, and the rate of drilling of wells increased by two and one-half times from 1921 to 1926. The number of wells drilled in the four years 1924 through 1927 was more than five times greater than the number drilled in the whole period before 1921."[61]

Yet for the most part this new drilling took place on lands controlled before 1917, and the government accused the companies of holding vast tracts of land in reserve (a charge frequently levied by nationalists against the majors in other Latin American countries). But the stepped-up drilling program did yield some results: in 1930 El Aguila (now controlled by Royal Dutch–Shell) found the aptly named Poza Rica deposits. Thanks to this strike Mexican oil production began to recover.[62]

The real decline in private investment occurred after 1926, as the companies turned to lower-cost, more promising and secure operations in Venezuela, Colombia, and the Middle East.[63] Why throw good money after bad, having concluded that Mexico was riskier?[64] In addition to government pressures, the foreign operators now had to face rising labor militancy, which increased their costs and did lead eventually to the long-feared expropriation.

Oil and Labor

The labor history of the industry is so complex that it merits a separate study. My aim here is to show labor's role in bringing about expropriation and in developing a nationalized industry.

The relations between Pemex and its workers are rooted in three main events, according to Powell: the revolution, Article 123 of the

1917 Constitution, and the 1931 Federal Labor Law that implemented this constitutional mandate.[65] Indeed, labor militancy in
the oil fields was rife during revolutionary days, particularly in
Tampico, where labor was in great demand. There the companies
competed with each other and with the power company, the tram-
ways, and the docks for labor. The Tampico workers took advantage
of their situation and became even more militant.[66] At first the companies opposed unionization and tried to prevent it by various
means, including the use of "white guards" to intimidate trade
union activists. But later they found it more expedient to organize
company unions themselves. In 1921 El Aguila was the first company to grant union recognition. By 1935 every company had a
union.[67]

Until expropriation the oil workers were organized in thirty-two
locals, by company and by function. Labor contracts, wages, and
working conditions varied from company to company and also between oil fields. There was so little uniformity that even within the
same refinery working contracts varied. The big strike at El Aguila
in 1934, however, did achieve positive results for the workers:
wages were equalized and regional commissions were established
to consider worker demands.[68]

The legal basis for labor rights and protection were specified in
Article 123 of the constitution. In one of its most progressive clauses, this article also defined something called "the conflict of economic order," a concept that was to become important for labor
relations. A conflict of economic order was said to arise when conditions changed so drastically as to make the continuation of a labor
contract socially undesirable. At that point, the conflict could be
presented to the Federal Board of Conciliation and Arbitration
(FBCA), which would maintain the status quo until three experts it
appointed had investigated a company's finances. Two panels, one
representing management and the other labor, had the right to protest the expert opinion. Then, based on this evidence, the FBCA
would decide.[69]

The inauguration of Lázaro Cárdenas in December 1934 and his
decidedly prolabor stand brought in a new era for the Mexican labor movement in general and for the oil workers in particular. By
then the oil workers had acquired considerable strength in terms of
numbers: in a total work force of 214,000 in manufacturing, 23,000

were in petroleum refining and production.[70] Furthermore, they were at the top of the Mexican salary scale in 1938.[71]

Soon after his inauguration, Cárdenas demonstrated his support for labor. After the oil unions won a series of strikes in early 1935, the Labor Department handed down several decisions regarding wage rates, wage uniformity, and contractual cases on the basis of the constitutional principle of equal pay for equal work.[72] Indeed, Cárdenas was acting to implement the National Revolutionary Party's Six-Year Plan, which called for the full enforcement of Articles 27 and 123 of the constitution. He also encouraged workers to unionize, leading to creation of the government-controlled Confederación de Trabajadores de México (CTM) in 1936.

In December 1935, all the independent oil locals united to form the Sindicato de Trabajadores Petroleros de la República Mexicana (STPRM), and shortly thereafter it affiliated with the CTM and came under government influence. The first general assembly of the STPRM met in July 1936 to draft the first collective contract covering all oil workers. But the workers' demands as spelled out in this contract were rejected by the companies. In November 1936, the threat of a general strike was averted by the government and by the onset of negotiations between labor and management.[73]

In a fourteen-point plan issued in February 1936, Cárdenas clearly spelled out his administration's attitude toward the labor movement, the industrialists, and labor-management conflict. Henceforth, the government would regulate and arbitrate labor relations. Worker demands would be considered in light of the companies' financial capacity to meet them. The government would deal with organized labor and not with splinter groups. Employers had the same right as workers to unite and form a common front. And, most important, Point 14 established that any industry that did not want to consider labor's demands could turn its installations over to the workers or to the government, since the closing of factories would not be in the interest of society as a whole.[74]

Nine months later the Chamber of Deputies passed the Expropriation Law, which was aimed at the employer who would close down his factory on the grounds of a lack of "economic capacity" to meet labor's demands. As it turned out in practice, this law cleared the way for the government to resolve labor-management conflict in the oil industry by expropriation. The companies claimed that this law was unconstitutional since it would benefit only one class and not the whole of society. This contradicted the Law of 1925, they said, which declared oil a public utility.[75]

All through 1936 and into early 1937, negotiations took place between STPRM and the companies, but with no agreement in sight a strike broke out in May. The government asked STPRM workers to return to work and to argue their case as a "conflict of economic order" to the FBCA. In December, basing its decision on evidence presented by the expert commission—composed of Under Secretary of the Treasury Efraín Buenrrostro, Under Secretary of the National Economy Mariano Moctezuma, and Jesús Silva Herzog, counselor to the Department of Finances and Public Credit—the FBCA made a settlement that moderated labor demands. The union requests would have more than doubled labor costs through increased basic wages and fringe benefits such as a savings fund, paid vacations, housing allowances, and a uniform pension system.[76] Although the settlement stopped short of granting all to labor, the companies were determined not to go along.[77] They appealed for an injunction against the award, on grounds they were financially unable to pay for what amounted to a 26-million-peso increase in their wage bill over the previous year's.[78]

The FBCA upheld the arbitrated settlement on March 1, 1938. But the companies insisted they could not pay, and the STPRM, in turn, requested that the work contracts be canceled, a position upheld by the FBCA. On March 18 all work ceased. In this impasse, the only possible means for Cárdenas to maintain political control and to appease the labor movement was to come out with a dramatic solution. This was nationalization, a move forced on Cárdenas by the unions and by his previously outspoken support for labor. Soon he announced on the radio that he was expropriating the oil companies because, with their "arrogant and rebellious attitude," they had not respected the decision of a sovereign nation.

The Aftermath of Expropriation

Having nationalized the oil industry, the Mexican state was faced with the complex task of organizing and managing an industry, the intermediate and long-range goals of which were not particularly well known. Even the objectives of expropriation had not been specified, because it was basically the outcome of labor dispute.

Here the contrast with the other three case studies in this book is striking. Mexican expropriation resulted from a national cause célèbre that united elites and masses; technocratic leaders set the agenda far less than they did in Argentina, Brazil, and Venezuela. Expropriation certainly did not result from a deliberate effort to fulfill a long-term economic development plan in which petroleum had been assigned a certain role. But with nationalist ideals of the revolution providing an overall policy umbrella, state control of this industry became a triumph of the revolution, in the words of Antonio Bermúdez "one of its fortunate achievements."[79]

The expropriation was applauded by the whole nation, which rallied behind the president. The people understood that the country was getting rid of foreign oil concerns that had long exploited Mexico's nonrenewable resources with the aim only of enriching themselves and with no concern for the welfare and development of the country. In the popular view, expropriation would change all this. As a public utility, oil would no longer be developed for profit, but rather in the best interest of the national economy. Implicit in this widespread support for nationalization, according to Powell, was the people's perception that "theirs was a semi-colonial country doomed to backwardness until it achieved economic independence and freedom from economic imperialism."[80]

Two main objectives for the new state industry were worked out. The first was to provide the nation with sufficient energy to promote its progress and development. The second was to turn the industry into a basic instrument of independent economic growth. Writing years later, after the 1973 oil shock, Bermúdez claimed that the first objective was achieved, although at a cost greater than could have been expected, but the second goal had not been reached.[81] He voiced this opinion when the oil industry was undergoing a severe crisis, and it typified the general disenchantment with Pemex's long-term performance. And this was the same man who had directed Pemex for twelve of its most successful years, from 1947 to 1958. That Pemex had not become the basic instrument of independent development was due, he concluded, to a series of errors on the part of both the central government and the state company.

These errors included: fixing excessively low retail prices for oil and derivatives; allowing profit-making to supercede public service; neglecting exploration; and the recurring indebtedness on a huge scale without long-term planning.[82] In the early 1970s, before the oil boom, there was indeed a general concern in official circles about the state of the oil industry. An indication of the crisis Pemex was undergoing is the fact that from 1966 to 1973 Mexico did not export any crude oil. It had to import crude, gasoline, diesel, kerosene, and other derivatives in unprecedented amounts from 1971 to 1973.[83] The origins of the preboom crisis of the 1970s go back to the days when the industry was nationalized.

At the outset, especially during the first two years after expropriation, the government faced a difficult mix of external and internal problems in trying to manage the industry. The foreign companies were actively hostile to Mexico. Having appealed to the Mexican courts, they complained to their governments and launched a publicity campaign against Mexico in the United States and abroad.[84] The oilmen also orchestrated an effective boycott both against Mexican oil in the international market and against the supply of equipment and services to Mexico by other foreign companies. This forced the government to engage in barter trade with the Axis powers. Attempts to sell what the companies called "pirated oil" to Latin American countries outside the marketing systems controlled by the majors were largely unsuccessful. The British severed diplomatic relations, but the United States with its Good Neighbor Policy accepted the right of Mexico to expropriate with fair and prompt compensation. The Sinclair interests began receiving payments in 1940, and settlements with other American companies were arranged in 1942. Royal Dutch–Shell was not paid off until 1947.[85]

At home complex political and administrative questions remained. How was the industry to be organized, and with what degree of control for the workers? If the state took over, what kind of relations were to be established with the STPRM? How was the government to cope with the lack of qualified technical personnel and to manage run-down installations that the departing companies had decapitalized? From the start it was a matter of national pride to overcome these problems and to demonstrate to the world that Mexico could manage its own oil industry.

The government had acquired some experience with managing an oil industry. As early as 1925, Calles had set up the Control de la Administración del Petróleo Nacional (CAPN) to compete with the private companies. CAPN had the advantage of being able to drill in federal reserves, outside previously granted concessions. In addition to producing, CAPN was enjoined to enter refining and to regulate the internal prices of petroleum products.[86] The concern with refining stemmed from the government's effort to compete effectively, and to do so CAPN had to become competent in an industrial activity as important as oil production itself.

In the early years, the companies exported mostly crude, but the government's policy of increasing export taxes on crude while setting lower taxes on exported refined products stimulated the building of refineries.[87] Before 1938 the government owned one refinery, the Bella Vista topping plant in Tampico. Expropriation brought five major refineries into government hands: four complete installations (Minatitlán, Ciudad Madero, Mata Redonda, and Arbol Grande); one topping plant (the Atzcapotzalco refinery), and a topping plant under construction (at Poza Rica). The topping plants excepted, all of these refineries were oriented to export markets.[88] And "in general," writes Powell, "the refining equipment inherited by Pemex was old, badly worn, inefficient and obsolescent."[89] A constant irritant to the Mexican ruling elite (and a concern of most Mexican scholars who have written on the subject)[90] was that oil was exported to the detriment of domestic consumption. That Mexican oil was developed to be consumed elsewhere and did not contribute to the development of Mexico's economy has become almost axiomatic.

However, it is worth remembering that the foreign companies did not altogether neglect the domestic market. In the early days of oil, the pioneers aimed to capture the *domestic* market. For example, Clay Pierce did not produce oil; his business was to supply and distribute oil products. Cowdray tried to compete with Pierce for this domestic market to the point of a price war roughly from 1908 to 1910. Doheny provided asphalt for street paving in Mexico, and most oil companies tried to become suppliers to the National Railways of Mexico.

Stimulated by high prices during World War I, however, the companies pushed exploration, and most of the oil they produced subsequently was exported. The 1920s brought a glut in the world oil supply, coupled with falling prices after 1929 that affected Mexican production, but just at the time when domestic consumption was rising. Indeed, a certain stability was achieved after the revolutionary turmoil and Mexico was initiating its own development process. Coal was scarce and of poor quality, but Mexico could rely on oil and consumption grew rapidly. By 1936, Wendell Gordon wrote, "Mexico consumed 16.89% of its production of heavy crude, 99.9 percent of the production of light crude, and 43.5 percent of the refined products produced."[91]

In view of the growing importance of the domestic market, plans were drawn up in 1933 for a semiprivate (mixed) enterprise to replace the Control de la Administración del Petróleo Nacional. Called Petróleos de México or Petromex, the aim of the new enterprise was, apart from refining which it took over from CAPN, to obtain a degree of control over the domestic market with the help of private capital. The expectation was that Petromex would start by fulfilling the government's oil needs, especially for the National Railways, and by training a technical staff to run the petroleum industry.

The extent to which Petromex was inspired by the Argentine example of YPF is a matter of speculation, for no direct evidence exists to confirm it. However, it is significant that in 1928 General Mosconi visited Mexico, where in a speech at the National University "he lashed out at the international oil companies, explained the progress of YPF, and urged the Latin American governments to cooperate to fight the 'exploitative trusts.' "[92] Solberg has stated that Cárdenas, who heard Mosconi speak, "espoused an ideology of petroleum nationalism similar to that of the Argentine general."[93]

Petromex followed the mixed company model of the type so often experimented with in Latin America in the desire to strike a balance between public control, private interests, and efficiency. It began operations in early 1934 and soon controlled 3 to 4 percent of national production. A major obstacle to the development of Petromex was that "only a small portion of the total authorized capital stock was ever subscribed by Mexican investors, and the concern did not really achieve the mixed status for which it was designed."[94] Since foreign capital could not invest and the government was short capital, Petromex was unable to compete effectively with the foreign oil companies and remained relatively undeveloped. Petromex did accomplish something, however. It built the Bella Vista topping plant at Tampico to serve the domestic market, and it established a distribution network while training Mexican personnel.[95] CAPN was liquidated in September 1934, and its assets transferred to Petromex. In 1937 Petromex was itself replaced by something called the Administración General del Petróleo in a move to place these assets under direct government control. The administracion was an executive agency, its director being appointed by the president.

Thus the government was not unfamiliar with the industry but the fact remained that "the industry to which Mexico fell heir by expropriation was pretty much a hodgepodge that needed extensive reorganization to form a rational national system. The administrative structure included somewhat less than a score of more or less parallel, independent organizations that had been built on a competitive basis."[97] There were also some good producing wells, whose rate of production compared favorably with California.[98] Mexico was well endowed to set up an internationally competitive oil industry, but first it needed an adequate organization, good labor relations, and cost-effective operations.

Initially, relations were cordial and cooperative between labor and government in its new role as manager. One day after expropriation, the Consejo Administrativo del Petróleo (CAP) was formed. CAP was headed by the Secretary of Economy Buenrrostro, aided by nine members appointed by the president—two from the Secretaría de Hacienda y Crédito Público, three from the Secretaría de la Economía Nacional, one from Administración del Petróleo Nacional, and three representatives from STPRM, the petroleum workers' union. CAP's purpose was to "manage the expropriated properties in such a way that an integrated petroleum industry could be managed and function for the welfare of the nation and of the workers of the oil industry."[99] As it turned out, the unions benefited most.

While the government was deciding how to organize the industry, STPRM, the only organized group with enough technical expertise to keep production up, took over most of the administrative positions. The CAP was soon relegated to an advisory rather than an executive role and the union exploited this weakness. Organized along company lines, which facilitated control by the locals' administrative councils, union members took charge and operated "with about the same administrative autonomy in technical matters as had the managements of the foreign companies."[100] Each local council was run by a general secretary, a representative of the Ministry of Labor, and a member of the Supervisory Councils (Consejos de Supervisión) who was designated by each of STPRM's thirty-two locals.[101]

CAP, in theory, was to coordinate these local administrative councils, but in practice the officers appointed by the federal government "found themselves helpless in the face of the Syndicate [STPRM] leaders, who controlled the local operations and rapidly established vested interests in their local control."[102] Since the union dictated appointments, jobs within the new oil organization multiplied and were assigned on a patronage basis to union officers.[103] Criticism of union control of the industry was not long in coming. Detractors cited the growth of jobs beyond the capacity to fund them as well as the incompetent performance of some union officials in high management positions.[104]

STPRM had hoped that the newly nationalized industry would be turned over to the workers, as Cárdenas had pledged in his election campaign and as he did do with the railways.[105] These hopes were dashed by the decree of July 20, 1938, which established the public enterprise called Petróleos Mexicanos, or Pemex, to take over the industrial properties and the public marketing corporation called the Distribuidora de Petróleos Mexicanos.[106] The Administración General del Petróleo Nacional would continue to run operations with Pemex.

The president appointed the Pemex management and the board, composed of two representatives from the Secretaría de Hacienda, three from the Secretaría de la Economía Nacional, one from the Administración General del Petróleo Nacional, and three from STPRM. In turn, the board appointed the company's general manager. Both the president and the board had to approve the budget. Any profits would go to the Secretaría de Hacienda.[107] The administración did make profits of 4 million pesos in 1939, but these were more than offset by "Pemex and the Distribuidora [which] had suffered a decline from profits of 15 million pesos in 1938 to losses of more than 21 million in 1939."[108] Increased operating costs linked mainly to labor overruns caused this debacle. In 1940, the industry ceased to pay taxes and the government had to grant Pemex 60 million pesos to liquidate its tax obligations.[109]

The union was threatened by the creation of the new public enterprises, and regretted that the new administrative setup was more akin to a private enterprise than to a government company. Although the aim of the new organizations was social betterment instead of profit making, they were supposed to be run in a businesslike manner and to be financially independent. Indeed, when Bermúdez became Pemex director-general later on he declared that Pemex ought to increase its capital and profits.[110] This aim was difficult to achieve. To use the industry as a resource to aid Mexican industrialization, oil was consumed internally and not exported massively as it had been under private ownership. And to facilitate industrialization, it became accepted policy to keep fuel costs low.

Prices under the foreign companies had been uniform and relatively stable. Although Powell asserts that "domestic prices of petroleum products were fixed by the government even before expropriation,"[111] this was only true for gasoline, which the government considered a commodity of prime necessity. (In the decree of September 13, 1935, the sales price of gasoline was fixed by the Department of National Economy.)[112] The government determined the internal prices for all oil products after expropriation. The pricing criteria were more social and political than economic. Two factors kept the prices of oil products low, the government's policy of encouraging consumption and the political strength of labor. A few price increases did take place during the initial years of Pemex's financial straits. But in fact the government's price policy subsidized consumption at the expense of the rational development of the oil industry.[113]

If this pricing policy hindered the public enterprise from reaching financial self-sufficiency, labor also complicated the development of efficient operations. STPRM had its own ideas about what the goals of this public enterprise should be, namely to care for and protect the workers. Soon the cooperation between government and labor gave way to open conflict. Not only did STPRM resent not having obtained full control of the industry, but it also insisted that provisions of the 1937 contract arbitration and award be put into effect.[114]

It is clear that the initial organization of the state-run petroleum industry into three distinct entities was inefficient. This encouraged overmanning and the duplication of tasks, causing higher running costs. Unmindful of the precarious situation of the newly nationalized industry, labor did not curb its demands for economic benefits and did not strive to obtain high levels of productivity. Finally, creating separate enterprises for the distribution of petroleum and for production and refining simply did not make administrative sense.

Roughly two years after expropriation the industry was in such disarray that a new organization with more central control was necessary. In January 1940, after meeting with managers of the three entities, Cárdenas called for an urgent reorganization of the industry and outlined its most pressing goals. The first and most important goal was to balance the budget. From gross receipts estimated at 235 million pesos in 1940, the units would have to cover their expenses and operations without other funding. The figure was not unrealistic—income had amounted to 233 million pesos in 1939—but to balance the budget with sums of this size required severe cuts, especially in labor costs.[115] To achieve this the president called for the trimming of excess personnel, an end to overtime, a salary review for the white collar workers, and reducing replacements. Cárdenas also called for a stepped-up drilling program, the modernization of refineries, and the improvement of all facilities including pipelines. He said the industry must also meet its tax obligations and repay the advances from the treasury to purchase new equipment. Finally, he decreed that 20 percent of export receipts would be set aside to create a fund to pay for the expropriated properties.[116]

To these cost-cutting measures the union reacted with disapproval, and a conflict "short of civil war" arose between government and labor over the terms of this reorganization. STPRM was no paper tiger (*sindicato charro*). According to Vicente Lombardo Toledano of the CTM and hardly a disinterested observer, the oil union could even count on the support of the ousted foreign companies, which "were prepared to swing their tremendous economic power behind any group potentially capable of destroying Cárdenas' power."[117] Silva Herzog, the financial expert, attacked this aspect of oil company activities, which he called subversive.[118]

Cárdenas succeeded in reorganizing the oil industry over union opposition by granting one of the STPRM's suggestions: to organize the industry into one entity. Thus on August 8, 1940, the Administración General and the Distribuidora transferred their assets to Pemex, making it a vertically integrated national petroleum company.[119]

Reorganization of the oil industry on a sounder basis took time, but the new structure facilitated more effective control by the political authorities. The basic legal instrument for executive control was the 1941 Petroleum Law, which declared the industry a public utility under federal jurisdiction. (Since 1925 it had been a public utility paying taxes to federal, state, and some municipal governments.) The federal authorities now had the sole right to tax it, to regulate its operations, and to authorize technical and administrative decisions concerning operations. Pemex would monopolize exploration and production, but the secretary of economy had the authority to grant concessions authorizing private participation in all other phases of the oil business.[120]

Pressures to produce more oil led the government to modify its stand on self-sufficiency in order to earn more revenues to pay for the expropriation. Increased production had priority in all of Cárdenas' reorganization plans. The capital requirements for stepped-up exploration and equipment led the government after World War II to encourage private investment. This change in policy was forced on the government by the urgent need to rehabilitate the industry, but it also coincided with *alemanismo*, a new rapproachment with private capital instituted by President Miguel Alemán. The possibility of United States government help for Mexican industrialization was discussed, and the government expressed interest in American technology to modernize the oil industry and the National Railways.[121]

Once again Mexican private capital did not want to enter this industry, although the government encouraged it to do so. (Recall the previous government efforts to promote participation in Petromex.) Opposition to foreign capital (especially by the STPRM) forced Pemex to contract with only small American companies that had little influence and economic power. These weaker companies were unable to fulfill their contracts because of equipment shortages during and immediately after the war.[122] Thus to obtain the necessary capital the government resorted to American loans.[123] As Powell concludes in his thorough study of the postexpropriation industry: "Undoubtedly, from a purely economic standpoint, Mexico could have better solved its petroleum problem by more extensive regulation and other steps short of expropriation." Despite setbacks, however, Pemex "became the largest enterprise, as well as the most important industry, public or private, within Mexico's mixed economy."[124] By becoming a symbol of economic independence and sovereignty Pemex achieved as well a most important social and political objective.

Summary and Conclusions

The Mexican petroleum industry was developed by foreign entrepreneurs, who brought in the first commercially successful oil
wells in Latin America. Native capitalists were not enthusiastic
then or later about investing in the industry, from Lord Cowdray's
lack of success in selling shares in El Aguila to the short-lived Petromex, a mixed corporation, or still later in the free-enterprise climate of post-1940 Mexico.

Controlled by foreigners, Mexican oil was produced mainly for
export, although their first concern had been to develop the domestic market. Mexico became a major exporter in the midst of revolution, but the tide of nationalism that was both cause and consequence of this armed struggle gave back to the nation control of its
mineral deposits under Article 27 of the reformist 1917 Constitution. Resentment against the companies by the Mexican public and
by the political elite was caused by the companies' profit-maximiz-
ing approach to oil development and the industry's lack of integration into the domestic economy. Their meddling in internal political affairs in the Peláez affair infuriated the nation. The populace
and the elites shared these views, producing consensus.

The early revolutionary governments did not contemplate expropriation seriously because they were overwhelmingly concerned
with establishing their own political legitimacy, both at home and
abroad. Their pressures on the companies were rather of a regulatory and a fiscal character. This changed when Cárdenas gave a new
impetus to the nationalist and social-reformist thrust of the revolution and implemented Articles 27 and 123 of the Constitution.
Meanwhile, a militant labor movement developed in a setting of
foreign economic penetration. Well organized under government
tutelage, the oil workers demanded benefits that the companies
found excessive. When the companies balked at an arbitrated settlement, this prolabor government had no alternative but to expropriate.

Labor militancy did not wane with the onset of state control, but
the government took strong action to wrest control from the oil
workers' union and to centralize it under one vertically integrated
public enterprise, Pemex. However, the broad aims assigned to
Pemex seemed incompatible. Labor thought the company existed to
benefit union members; racketeering and corruption became en-

demic. Then again, Pemex was called upon to produce more oil at low, subsidized prices to benefit consumers and to support industrialization. Deprived of substantial earning power through market pricing, it was nonetheless supposed to balance its budget and be a self-financing enterprise. After 1938 the government faced severe social and economic problems in managing the industry. But the state was able to cope, and Pemex with all its weaknesses and vices did emerge as a viable public enterprise, the symbol also of national independence, something that is being tested in the oil boom and bust of our times.

Notes

Abbreviations

Anderson Diary	Diary of Chandler P. Anderson, Manuscript Division, Library of Congress
DS	State Department Archives, Washington, D.C., Record Group 59
Fall Papers	Papers of Albert Bacon Fall, Huntington Library, San Marino, California
FO	Public Record Office, London, British Foreign Office Files
HASPA	Historical Archives of S. Pearson and Son, Science Museum, London

1. *Oil and Gas Journal*, Dec. 28, 1980; David Ronfeldt, Richard Nehring, and Arturo Gándara, *Mexico's Petroleum and U.S. Policy: Implications for the 1980s* (Santa Monica, Calif., 1980), p. 3

2. It may be of interest to note in passing some parallel developments pointing in the same outward-looking direction, such as the more ambitious and structured foreign policy of the José López Portillo regime (the World Energy plan presented at the United Nations and the San José Agreement, jointly sponsored with Venezuela) as well as the surprisingly vigorous debate on entry to GATT.

3. René Villarreal, "El petróleo como instrumento de desarrollo," *Cuadernos sobre Prospectiva Energética* 1 (working papers) (Mexico City, April 1980). However, this argument overlooks, and policy making probably did overlook, the fact that a sudden massive export boom may raise and keep up the value of the local currency, negatively affecting traditional export-oriented (or import-competing) sectors and activities. See: W. Max Corden and J. Peter Neary, "Booming Sector and De-industrialisation in a Small Open Economy," *Economic Journal* 92 (1982):825–48.

4. Over the period 1914–18, Britain imported 72.2 million barrels from the United States and 7.3 million directly from Mexico, but the United States imported 108.9 m. from Mexico over the same period. For greater detail on oil trade figures between these countries see Esperanza Durán, "El petróleo mexicano en la Primera Guerra Mundial," in *La energía en México: Ensayos sobre el pasado y el presente,* ed. Miguel S. Wionczek (Mexico City, 1982) pp. 53–73. Also, DS 812.6363/411, Mark Requa to Frank Polk, July 10, 1920.

5. Ronfeldt et al., *Mexico's Petroleum and U.S. Policy,* p. 51.

6. Ralph W. Hidy and Muriel E. Hidy, *Pioneering in Big Business, 1882–1911,* vol. 1 of *History of Standard Oil Company (New Jersey)* (New York, 1955), pp. 448f., 646.

7. U.S. Congress, Document No. 285, 66th Congress, *Investigation of Mexican Affairs,* 2 vols. (Washington, D.C., 1920), 1:215–16; 222–28, 227–43.

8. Cowdray's biographers are J. A. Spender, *Weetman Pearson, First Viscount Cowdray, 1856–1927* (London, 1930), and Desmond Young, *Member for Mexico* (London, 1963); both are laudatory but informative works.

9. Robert Keith Middlemas, *The Master Builders* (London, 1963), p. 219; Cleona Lewis, *America's Stake in International Investment* (Washington, D.C., 1938), p. 222n.

10. J. Richard Powell, *The Mexican Petroleum Industry, 1938–1950* (Berkeley and Los Angeles, 1956), pp. 58ff. Lorenzo Meyer, *México y los Estados Unidos en el conflicto petrolero, 1917–1942,* rev. ed. (Mexico City, 1972), p. 19. The contribution of the independent Mexican-owned oil companies to the total production of oil from the beginning of the industry until its nationalization fluctuated between 1 and 5 percent.

11. Lewis, *America's Stake,* pp. 262, 612f.

12. Mira Wilkins, *The Maturing of International Enterprise: American Business abroad from 1914 to 1970* (Cambridge, Mass., 1974), p. 36. Table 5, based on Miguel Manterola, *La industria del petróleo en México* (Mexico City, 1938), p. 387, shows a higher figure (16.7 million pesos) for the same year.

13. On oil legislation see: Mexico, Secretaría de Industria, Comercio y Trabajo, *Documentos relacionados con la legislación petrolera mexicana,* 2 vols. (México, 1919–22); Manterola, *La industria del petróleo en México* 1:1–27; Merrill Rippy, *Oil and the Mexican Revolution* (Leiden, 1972), chs. 1–4.

14. "The Nation has had, and has, the right to convey title [on lands, etc.] to private persons, so establishing private property" (my emphasis) (Robert Freeman Smith, *The United States and Revolutionary Nationalism in Mexico, 1916–1932* [Chicago, 1972], p. 267, appendix).

15. Robert Glass Cleland, ed., *The Mexican Year Book, 1920–21* (Los Angeles, 1922), p. 299; Meyer, *México y los Estados Unidos,* p. 34.

16. Antonio J. Bermúdez, *The Mexican National Petroleum Industry* (Stanford, 1963), p. 4.

17. DS 812.6363/124, Clarence Miller (American consul in Tampico) to secretary of state, July 2, 1914.

18. Harold E. Davis, "Mexican Petroleum Taxes," *Hispanic American Historical Review* 12, no. 4 (1932):406.

19. Rippy, *Oil and the Mexican Revolution*, pp. 30f.

20. Cleland, *Mexican Year Book*, p. 299.

21. José López-Portillo y Weber, *El petróleo de México: Su importancia, sus problemas* (Mexico City, 1975), pp. 35–38.

22. Friedrich Katz, *The Secret War in Mexico: Europe, the United States and the Mexican Revolution* trans. Loren Goldner (Chicago, 1981), p. 133.

23. José Domingo Lavín, *Petróleo: Pasado, presente y futuro de una industria mexicana* (Mexico City, 1950), p. 107.

24. FO 371, vol. 2395, fol. 6023, Cowdray to Nicolson, Jan. 16, 1915. See also correspondence between American oil companies and Mexican authorities in U.S. Department of State, *Papers Relating to the Foreign Relations of the U.S., 1915* (Washington, D.C.), pp. 870–91.

25. For Pastor Rouaix's proposal to cancel concessions granted to El Aguila by the Díaz government see DS 812.6363/214, Kellogg to Canova; New York *World*, Feb. 9, 1916; Fall Papers, File T6, Fall to F. K. Lane, Jan. 12, 1917.

26. In effect the Calvo Clause; see *Foreign Relations, 1917*, p. 1059.

27. Decree in *Foreign Relations, 1918*, pp. 713f.

28. Smith, *Revolutionary Nationalism*, pp. 117–19.

29. Ibid.

30. Anderson Diary, March 10, 1917; FO 371, vol. 2402, fol. 145052, minute by M. de Bunsen, Oct. 7, 1915; FO 371, vol. 2402, fol. 147789, Cowdray to M. de Bunsen, Oct. 10, 1915; FO 371, vol. 2959, fol. 40734, minutes, Feb. 2, 1917; HASPS, box A-3, Body to Cowdray, memo, April 29, 1917.

31. George Philip, *Oil and Politics in Latin America: Nationalist Movements and State Companies* (Cambridge and New York, 1982), pp. 317.

32. Government of Mexico, *Mexico's Oil: A Compilation of Official Documents in the Conflict of Economic Order in the Petroleum Industry, with an Introduction Summarizing Its Causes and Consequences* (Mexico City, 1940), p. 879.

33. Beatriz Rojas, "Chronique et sociologie de la Révolution Mexicaine, 1910–1920: Le Groupe Carrera Torres/Cedillo" (Thèse de Troisième Cycle) Université de Montpellier, 1976), p. 133.

34. John Womack, Jr., *Zapata and the Mexican Revolution* (New York, 1968), pp. 423, 427, 566f.

35. Charles C. Cumberland, *Mexican Revolution, The Constitutionalist Years* (Austin, 1972), p. 251; Jesús Silva Herzog, *Trayectoria ideológica de la Revolución Mexicana* (Mexico City, 1976), p. 128; Rojas, "Chronique et Sociologie de la Révolution Mexicaine," p. 132. Rojas had access to the closed archives of the Mexican Ministry of Defense, where Peláez's bibliographical file is kept.

36. FO 371, vol. 2399, fol. 54101, minute; vol. 2399, Spring Rice to FO, tel. 513, May 4, 1915; *Foreign Relations, 1915*, pp. 870–91.

37. According to Josephus Daniels, Peláez received $35,000 a year from the oil companies (Edmund David Cronon, ed., *The Cabinet Diaries of Josephus Daniels, 1913–21* [Lincoln, Neb., 1963], p. 214). Harold Walker reported Peláez received $50,000 per month: $20,000 from Doheny, an equal amount from Lord Cowdray, and the rest from smaller companies (Smith, *Revolutionary Nationalism*, p. 104n). The League of Free Nations Association estimated Peláez received $200,000 monthly (Meyer, *México y los Estados Unidos*, p. 100). The Los Angeles *Daily* (May 26, 1917) reported that Peláez was receiving $80,000 per month.

38. Smith, *Revolutionary Nationalism*, pp. 103f.

39. Anderson Diary, March 10 and 11, 1917; FO 371, vol. 2959, fol. 60106, desp. 137, March 1, 1917; Dennis J. O'Brien, "Petróleo e intervención—Las relaciones entre los Estados Unidos y México 1917–1918," *Historia Mexicana* 27 (1977):103–140, esp. 117f.

40. HASPS, A3, H. Carr to Cowdray, March 14, 1917; FO 371, vol. 2959, fol. 52269, S. Pearson and Son to R. Sperling, March 12, 1917. The reasons Cowdray refused to supply Peláez with arms were that he did not want to become "the catspaw of the American oil companies" and that supplying Peláez with arms could not be done without the knowledge of the Carranza government, from which Cowdray feared retaliations.

41. London *Times*, Dec. 12, 1917; *South American Journal*, Oct. 20, 1917 *The Financier*, Dec. 22, 1917; *Financial and Commercial Chronicle*, Nov. 17, 1917|—all in Newspaper Files of the Council of Foreign Bondholders, Guildhall Library, London, microfilm, reel 13—and New York *Herald*, Nov. 20, 1917.

42. The embargo was imposed on the exports of these commodities by the United States after her entry into the war. This embargo was greatly resented in Mexico, which was heavily dependent on this commerce. Carranza sent Luis Cabrera and Alberto Pani to Washington to demand the lifting of the embargo, but to no avail. See Thomas A. Bailey, *The Policy of the U.S. toward the Neutrals, 1917–1918* (Baltimore, 1942), p. 318; Francis William O'Brien, ed., *The Hoover-Wilson Wartime Correspondence*, (Ames, Iowa, 1974), pp. 33, 41f. Conclusive evidence as to Carranza's motives is lacking. Carranza's papers at Condumex (Mexico City) do not contain relevant information on this topic, which should be explored.

43. John W. F. Dulles, *Yesterday in Mexico: A Chronicle of the Revolution, 1919–1936* (Austin, 1961), p. 53.

44. Linda B. Hall, *Alvaro Obregón, Power and Revolution in Mexico, 1911–1920* (College Station, Texas, 1981), p. 242.

45. John G. McLean and Robert W. Haigh, *The Growth of Integrated Oil Companies* (Boston, 1954; reprint, Boston, 1970), pp. 84f.

46. Philip, *Oil and Politics in Latin America*, pp. 43f.

47. Practically, "viable nationhood" meant that Mexico had to qualify for receipt of foreign loans by settling the international claims brought about by foreign interests during the revolution.

48. Dulles, *Yesterday in Mexico*, pp. 156–57; Ernest Gruening, *Mexico and Its Heritage* (New York, 1928; reprint, Westport, Conn., 1968), p. 597; U.S. Congress, *Investigation of Mexican Affairs*, 1:62–67.

49. Fall Papers, box 106, file 1, confidential report to Senator Fall, Sept. 29, 1921. This treaty, proposed by the United States government, was much discussed in the diplomatic exchanges between the two governments from 1921 to 1923.

50. Government of Mexico, *The True Facts about the Expropriation of the Oil Companies' Properties in Mexico* (Mexico City, 1940), p. 40.

51. Ibid., p. 41.

52. Davis, "Mexican Petroleum Taxes," p. 412.

53. Ibid., pp. 413–14.

54. Fall Papers, box 106, folder 1, Walter Teagle et al. to Charles E. Hughes, New York, Aug. 18, 1921.

55. N. Stephan Kane, "Bankers and Diplomats: The Diplomacy of The Dollar in Mexico, 1921–1924," *Business History Review* 47, no. 3 (Autumn 1973):334–52.

56. Meyer, *México y los Estados Unidos*, pp. 177f.

57. Ibid., pp. 208, 275.

58. It is important to bear in mind that the oil companies were much worried about this action taken by United States representatives at the Bucareli Conferences, as the establishment of a precedent in Mexico by which they agreed to be despoiled of their property rights might jeopardize their other oil investments in Latin America.

59. Manterola, *Industria del petróleo en México*, p. 25.

60. Fall Papers, memorandum from H. Foster Bain (director, U.S. Bureau of Mines) to A. B. Fall (secretary of the interior), Washington, D.C., Nov. 10, 1921; *The Economist* (Oct. 15, 1921).

61. J. Richard Powell, *The Mexican Petroleum Industry, 1938–1950* (Berkeley and Los Angeles, 1954; reprint, New York, 1972), p. 15 and table 2, p. 209.

62. Ibid., pp. 55f.

63. Edwin Lieuwen asserts that oil development in Venezuela in the interwar period was "largely a reaction of Shell and Standard to growing nationalism and labor troubles in Mexico" (pp. 192–93, this volume).

64. Jonathan C. Brown, p. 27, this volume.

65. Powell, *Mexican Petroleum Industry*, p. 124.

66. The most comprehensive history of the industrial workers in Tampico (particularly the oil workers) is Lief Adleson, "Historia social de los obreros industriales de Tampico, 1906–1919" (Ph.D. diss., Dept. of History, Colegio

186   Esperanza Durán

de México, 1982); see also Marjorie Ruth Clark, *Organized Labor in Mexico* (New York, 1934), p. 81.

67. Philip, *Oil and Politics in Latin America*, p. 212; Jesús Silva Herzog, *Petróleo mexicano: Historia de un problema* (Mexico City, 1941), p. 99.

68. Joe C. Ashby, *Organized Labor and the Mexican Revolution under Cárdenas* (Chapel Hill, 1967), p. 196.

69. Ibid., pp. 63f.

70. According to the 1935 Industrial Census there were 16,000 laborers and employees in the oil fields and 7,000 in refineries. The Petroleum Bureau of the National Economy on the other hand estimated 18,000 in 1936, but this figure probably includes only permanent employees. See Government of Mexico, *Mexico's Oil*, p. 198. The best available data on number of workers in petroleum production after 1938 is Antonio J. Bermúdez, *The Mexican National Petroleum Industry: A Case Study in Nationalization*, (Stanford, 1963), p. 249, table VIII-1. Manterola's figure for the total number of workers in the oil industry is 15,255 for 1935 (*Industria del petróleo en México*, p. 68).

71. Ashby, *Organized Labor under Cárdenas*, p. 195f.

72. The president's message to the nation, March 18, 1938, in Government of Mexico, *Mexico's Oil*, p. 878.

73. Silva Herzog, *Petróleo mexicano*, p. 99f.

74. Ashby, *Organized Labor under Cárdenas*, pp. 212f.

75. Francisco López Aparicio, in *El Movimiento Obrero en México* (Mexico, 1952), p. 217, claims that this law was illegal, since it contradicted Article 27 and the individual guarantees granted by the constitution.

76. For the specific improvements in working conditions provided by the award see J. Richard Powell, "Labor Problems in the Mexican Petroleum Industry 1938–1950," *Inter-American Economic Affairs* 6, no. 2 (Autumn 1952), pp. 5–8.

77. This reflected a concern on the part of the oil companies that, by allowing oil-producing countries to examine and regulate their finances, they would be establishing a dangerous precedent in Latin America.

78. Bermúdez, *Mexican National Petroleum Industry*, p. 14; Silva Herzog, *Petróleo mexicano*, p. 110.

79. Bermúdez, *Mexican National Petroleum Industry*, p. 19.

80. Powell, *Mexican Petroleum Industry*, pp. 25f.

81. Antonio J. Bermúdez, *La política petrolera mexicana* (Mexico City, 1976), pp. 13–19, 35, 52.

82. Ibid., pp. 52–61.

83. Petróleos Mexicanos, *Anuario Estadístico 1981* (Mexico City, n.d.), pp. 117, 123.

84. See for instance the booklets published by Standard Oil between 1939 and 1940: *Whose Oil Is It?*, *The Underlying Facts in the Seizure of the Oil Properties*, and *Mexico's Inability to Pay*.

85. The topic of foreign reactions to the act of expropriation will not be dealt with here. On this see: J. Basurto, *El conflicto internacional en torno al petróleo de México* (Mexico City, 1976); Meyer, *México y los Estados Unidos*, pp. 359–442.

86. Powell, *Mexican Petroleum Industry*, p. 35; Bermúdez, *Mexican National Petroleum Industry*, pp. 9f.

87. Manterola, *Industria del petróleo en México*, pp. 59–62; Mexico, Secretaría del Patrimonio Nacional, *El Petróleo de México* (Mexico City, 1963); Gustavo Ortega, *La industria petrolera mexicana: Sus antecedentes y su estado actual* (Mexico City, 1936), p. 14.

88. Powell, *Mexican Petroleum Industry*, pp. 71–73.

89. Ibid., p. 73.

90. Meyer, *México y los Estados Unidos*, p. 37; López-Portillo y Weber, *El petróleo de México*, pp. 260f.; Manterola, *Industria del petróleo en México*, pp. 182–84; Ortega, *La industria petrolera mexicana*, p. 16. Meyer has stated that "the [oil] producing companies had so neglected the domestic market that in 1928 10 percent of the fuel consumed had to be imported, and this proportion increased to 14 percent in 1932" (*El conflicto social y los gobiernos del Maximato* [Mexico City, 1978], p. 53).

91. Wendell C. Gordon, *The Expropriation of Foreign-Owned Property in Mexico* (Westport, Conn., 1941; reprint, 1975), p. 80.

92. Carl E. Solberg, *Oil and Nationalism in Argentina: A History* (Stanford, 1979), p. 131.

93. Ibid., p. 182.

94. George King Lewis, "An Analysis of the Institutional Status and Role of the Petroleum Industry in Mexico's Evolving System of Political Economy" (Ph.D. diss., University of Texas, Austin, 1959), p. 148.

95. Powell, *Mexican Petroleum Industry*, pp. 26, 35; Ashby, *Organized Labor under Cárdenas*, p. 247.

96. Bermúdez, *Mexican National Petroleum Industry*, p. 10.

97. Powell, *Mexican Petroleum Industry*, p. 36.

98. Ibid.

99. *Diario Oficial* (Mexico City), March 30, 1938.

100. Lewis, "Institutional Status and Role of the Petroleum Industry," pp. 164–65.

101. Ibid., p. 165.

102. Powell, "Labor Problems in the Mexican Petroleum Industry," p. 9.

103. The secretary-general of the STPRM's general executive committee became head of personnel in Pemex, the secretary-general of section 4 of the STPRM was appointed general manager of the oil refinery at Atzcapotzalco; a leader of section 5 became chief of domestic sales. "Excepting the three managers . . . and possibly ten other high officials appointed by the government the union controlled the important management posts" (Lewis, "Institutional Status and Role of the Petroleum Industry in Mexico," p. 167).

104. Jesús Silva Herzog—one of the commissioners in the FBCA investigation of the conflict of economic order and later appointed general manager of the Distribuidora de Petróleos—deplored the rejection of his proposals to do away with superfluous positions created by the union to introduce economy and efficiency in his organization. Moreover, he said that some of the union officials in charge of executive positions were incompetent (*Petróleo mexicano*, pp. 237–39).

105. Gordon, *Expropriation of Foreign-Owned Property*, p. 89.

106. It is well known that several people close to Cárdenas, such as Francisco Múgica and Vicente Lombardo Toledano, advocated and supported the expropriation of the foreign oil sector. But just where the idea originated of actually organizing the newly nationalized sector into a state company remains obscure because the information to document it seems to be lacking.

107. *Diario Oficial*, July 20, 1938.

108. Powell, *Mexican Petroleum Industry*, p. 139.

109. Ibid., p. 161.

110. *El Nacional*, Dec. 1946, in ibid., p. 38.

111. Powell, *Mexican Petroleum Industry*, p. 106.

112. Government of Mexico, *Mexico's Oil*, p. 174; Manterola, *Industria del petróleo en México*, p. 221.

113. Powell, *Mexican Petroleum Industry*, pp. 107–11.

114. After expropriation the oil industry was not in the best financial conditions to grant labor's demands, but since just these demands had led to nationalization, the directors of Pemex offered to gradually implement the FBCA award. According to Powell, this was a hasty decision in view of the state of the industry. In July 1938 the board of directors of Pemex approved the increased wages of the award with certain discounts, but the new pay scales had to be granted to an alarmingly increased number of workers. In fact, the total labor bill according to Silva Herzog (*Petróleo mexicano*, p. 241) increased 89.99 percent from April 1938 to October 1939.

115. Powell, *Mexican Petroleum Industry*, p. 131.

116. Silva Herzog, *Petróleo mexicano*, pp. 248–53.

117. Lewis, "Institutional Status and Role of the Petroleum Industry in Mexico," p. 187. Excerpts from Lombardo Toledano's speech are in Silva Herzog, *Petróleo mexicano*, pp. 275f.

118. Silva Herzog, *Petróleo mexicano*, p. 274.

119. *Diario Oficial*, Aug. 9, 1940.

120. *Diario Oficial*, June 18, 1941.

121. Luis Medina, *Civilismo y modernización del autoritarismo* (Mexico City, 1979), pp. 82f.

122. Powell, *Mexican Petroleum Industry*, pp. 48–49.

123. Ibid., pp. 49, 168–69.

124. Ibid., p. 197. For an assessment of Pemex in recent years, consult George W. Grayson, *The Politics of Mexican Oil* (Pittsburgh, 1980).

Edwin Lieuwen

The Politics of Energy in Venezuela

Petroleum is the axis of the Venezuelan economy. It concentrates overwhelming power in the state and provides the chief props for domestic industry and foreign commerce. Although the oil industry employs only 2 percent of the labor force, it accounts for nearly 40 percent of the gross national product and provides the government with two-thirds of its revenues. In addition, oil exports provide the nation with 95 percent of its foreign exchange requirements.

Thus, petroleum accounts for Venezuela's high per capita income—the highest in Latin America ($4220 in 1981, with Argentina second at $2560), her financial solvency (until the post-1980 crisis), and the dominant foreign trade sector in her overall economy. Oil income has enabled the government to invest heavily in economic development and social welfare. Oil exports have made the currency strong and have permitted an extraordinarily large volume of imports of foodstuffs, manufactures, and capital goods.

Petroleum has been a mixed blessing for Venezuela, however. It has made her economy lopsided, precariously dependent upon international markets, and extremely sensitive to foreign political and economic changes. Moreover, the oil industry's demand for goods and services and its heavy payments to the state have raised domestic prices. Petrodollars have steadily, until very recently, appreciated the bolivar, so that for the nonoil economic sectors, it is cheap to import and expensive to export. As a result, Venezuelan industry and agriculture have been unable to compete with foreign suppliers in the domestic market without substantial tariff protection (which also raises consumer prices) and cannot compete in foreign markets without government subsidies.

The benefits from oil have not been widely distributed. High do-

190   Edwin Lieuwen

5. Oil nationalization poster, Venezuela, 1976. Courtesy James Breedlove.

mestic prices constitute a heavy drain on the limited purchasing power of the general population. Since profits are concentrated in too few hands, private savings, lacking domestic outlet, tend to go into domestic real estate or investment abroad. Thus, not only does the nation's great oil wealth encourage inflation, but it also does little to raise the living levels of the masses. For the past half-century, the state has had ambitious programs to "sembrar el petróleo," that is, to use the oil revenues to promote economic development and social welfare, but there is convincing evidence that they have exacerbated the already deep inequalities in Venezuelan society.

The petrodollar boom of the past decade has been as much a curse as a blessing. By providing the state overwhelming economic power it has distorted progressive political processes and normal political development, weakening the party system. Moreover, the increasingly exclusive role of the state in economic development has crowded out the private sector. And growing government waste, inefficiency, and corruption have been by-products of the oil boom.

Since 1979, the overdependency upon petrodollars has caused the nation severe economic problems. The weakening of the international oil market has produced a financial crisis and a recession in Venezuela. Excessive government reliance on dwindling oil exports and falling prices has brought an unfavorable trade balance, heavy foreign borrowing, a depletion of official reserves, and a sharp rise in the public debt. The government currently owes foreign banks an estimated $25 billion. Interest and principal payments falling due in 1984 total $13 billion while total exports are projected at only $14 billion. Hence, without International Monetary Fund help Venezuela cannot meet its current obligations.

The outlook for 1984 and beyond remains clouded by the weak international oil market. Even if oil prices improve, Venezuela must soon make a heavy, costly, protracted investment to increase production. Not only has rapidly rising domestic consumption reduced the amount of oil available for exports, but proven reserves, adequate for only 15 to 20 years at the present production rate, are not expanding rapidly enough to assure a viable long-range future for the Venezuelan oil industry. Furthermore, expanded production will be mainly in heavier oil, which is less valuable and more expensive to produce.

Introduction

Venezuela's energy policy is unique in Latin America. Her policy problems were quite unlike those experienced by any other nation in the area. Her concern, unlike that of Brazil and Argentina, for example, was not one of supplying domestic energy needs, which until very recently never required more than a small fraction of production. Rather, Venezuelan policy makers' principal problem was the sale of her petroleum in foreign markets, mainly the United States.

Despite the fact that for more than forty years (from 1928 to 1969) Venezuela was the world's leading exporter of petroleum—from 1926 to 1947 her production exceeded that of the entire Middle East combined—her government was unable to exert control over the market.[1] This was because the Venezuelan oil industry until 1976 belonged almost exclusively to the world's two largest oil corporations, Standard Oil of New Jersey (Exxon) and Royal Dutch–Shell (Shell), whose capital, oil resources, and diplomatic influence dwarfed those of the Venezuelan nation.

For the first half-century of its petroleum development, from 1920 to 1970, Venezuelan energy policy was primarily a dependent object, rather than the autonomous subject, of history. Though the nation owned the petroleum, the concessionaires owned the capital, the drilling equipment, the surface installations, the managers, and the technicians necessary to extract the crude. Not only did Standard and Shell control production but also they exercised dominion over transportation, processing, and distribution. These were vertically integrated businesses. They owned the pipelines and pumping stations that moved the oil from wells to their portside storage tanks. They owned the tankers that carried it from Venezuela to their refineries in Aruba, Curaçao, and the Atlantic coast of the United States. And they owned the distribution centers—the service stations that sold the refined products to consumers throughout the world.[2]

Although Venezuela was long the biggest upstream crude supplier for Standard and Shell's downstream marketing outlets, her government was unable to control either output or prices, for these companies owned alternate sources of crude, in Mexico and Asia between the two world wars and in the Middle East and North Africa thereafter. In fact, the development of the Venezuelan petroleum between the two world wars was largely a reaction of Shell and Standard to growing nationalism and labor troubles in Mexico. As their profits were threatened and their flow from older wells declined in the early 1920s, they began to disinvest in the Vera Cruz and Tampico areas and to move their operation to Lake Maracaibo. Thus, Mexico, the world's leading petroleum producer in the early 1920s, was replaced by Venezuela in this role in 1928.[3] Similarly, Venezuela's post-World War II nationalism and labor troubles prompted Shell and Exxon to disinvest there and to shift their emphasis to the Middle East, where taxes were lower, labor was more docile, new wells were bigger producers, and profits, accordingly, were better.

Not until 1970 were Venezuelan authorities able to exercise significant control over price and output. The reason was that oil was in surplus. Not until demand exceeded supply were the sellers able to break the century-long marketing monopoly of the producing companies and make their nationalistic policies effective. The rapid rise in the 1960s of the so-called independents to a position where they were able to compete with the hitherto dominant Seven Sisters cartel abetted this trend.[4]

And yet, even though Venezuela was unable, until the 1970s to implement effectively her nationalistic policy initiatives—which began in 1936 and grew in intensity thereafter—a study of her policy history is highly instructive and of world-wide contemporary significance. Venezuela, ever since World War II, led the campaign of the oil-producing countries to exert greater control over output and prices. Her tax legislation of the 1940s and 1950s spread to all other oil-exporting countries, and her creation and subsequent promotion of the Organization of Petroleum Exporting Countries (OPEC) in the 1960s was a critical factor in the exporting nations' wresting control of their oil industries from the grip of the Seven Sisters cartel. This, in turn led, during the 1970s, to the greatest transfer of wealth the world has ever known. Thus Venezuelan policy makers, especially President Rómulo Betancourt and Mining Minister Juan Pablo Pérez Alfonzo, patriotic heroes in their own country, were the principal architects of the 1970s world energy crisis. Venezuela's key role in bringing about this crisis can be clearly demonstrated from a survey of her petroleum policy experience over the past sixty years.

The Gómez Era: Rise of the Industry (1907–35)

Military dictators were the only presidents who ever granted oil concessions in Venezuela. The first concessions were granted in 1907 by Gen. Cipriano Castro to four close friends on the lands surrounding Lake Maracaibo, where petroleum had seeped through the soil for centuries, The concessionaires were granted fifty-year titles over huge areas and were required to pay about one dollar per ton in royalties. Castro's dictatorial successor, Gen. Juan Vicente Gómez (1908–35), granted additional concessions, with thirty-year titles and 5 percent royalties to favored Venezuelans in the years 1909–12. Nearly all of these 1907–12 concessions were sold by the owners in 1913 to Shell, which brought in the first commercial producer just east of Lake Maracaibo in February 1914.[5] Equipment and transport shortages delayed development during World War I, but that war conclusively demonstrated the strategic importance of oil and set off a worldwide scramble among the victors for control of this precious resource.

The British government's and Shell's aggressive postwar acquisitions policy was countered in 1919 by similar policies of the Woodrow Wilson administration and the big United States oil companies. By early 1920 the latter's agents were combing Venezuela for concessions to rival those of the British. In an attempt to capitalize on the foreign competition for Venezuela's petroleum, Gómez' Development Minister, Gumersindo Torres, drafted the 1918 oil law, which maintained thirty-year titles but reduced the size of new concessions and raised royalties to 15 percent. Late in 1921 United States Minister Preston McGoodwin and the American companies protested these harsher terms to Gómez and vowed not to invest unless better inducements were made. Gómez replied: "You know about oil. You write the laws. We are amateurs in this area."[6] Accordingly, the recalcitrant Torres was fired, and oil company lawyers drafted their own law, which was dutifully passed by Gómez' puppet Congress on June 13, 1922. The United States companies operated under this basic law until 1943. British companies continued to operate under the more favorable terms of the 1907–12 concessions.

The 1922 law gave the president sole authority to grant concessions. It provided for 10,000-hectare three-year exploration leases, from which the concessionaire was obliged to select alternate checkerboard 200-hectare exploitation parcels, the other half being returned to Venezuela as national reserves. Titles were for forty years; royalties were set at 10 percent (7½ percent for underwater parcels in Lake Maracaibo); customs exemptions were granted for industry-related imports.[7] "The best petroleum law in the world— for the companies," complained the embittered Torres, for nowhere did they have such favorable terms, and no nation got so little in return.[8]

General Gómez, however, profited handsomely. He granted concessions exclusively to his cronies, who in turn resold them to United States companies. The law thus was converted into a device for siphoning public money into private pockets. Sales of the national reserve parcels produced even greater private gain, for when producing wells came in on the concessionaires' parcels the value of the adjacent parcels was greatly enhanced. Under the law, these national reserves belonged to the national oil company, the Compañía Venezolana de Petróleo. This, however, sold these parcels at low prices exclusively to Gómez' friends, who in turn sold them at high prices to the oil companies. In this manner the dictator and his cronies bilked the treasury of additional millions. When it came to extracting resources from big oil, Gómez was far from helpless.[9]

1923–30 was the first boom era for the Venezuelan petroleum industry. Production was insignificant until December 14, 1922, when Shell's drillers struck Barrosa No. 2, a gusher that blew away the derrick and shot a column of oil two hundred feet into the air. The oil flowed wild at over 100,000 barrels per day (b/d)—"the most productive well in the world," reported the *New York Times*.[10] This discovery opened up the famous Bolívar Coastal Fields on the eastern edge of Lake Maracaibo, from which nearly two-thirds of Venezuela's production was to come. A frantic offsetting scramble took place among concessionaires: Shell on the land, Gulf on the shoreline, and Standard Oil of Indiana in the lake. Production rose exponentially from two million barrels in 1922 to 137 million in 1929. The huge per-well production volumes, the low transport costs (the company's shallow draft tankers picked up the adjacent production at lakeside storage terminals), the low political risk factors (an all-powerful dictator and a docile labor force), and "the best petroleum law in the world" combined to give Venezuela an international cost advantage that elevated her in 1928 to the world's leading exporter and the world's second-largest producer (after the United States), a position she forfeited to Russia during the 1930s.[11]

The royalties enriched Gómez and his favorites and were used to strengthen the army and thereby solidify the despotism. The Venezuelan nation also had its first taste of what contemporary third world scholars refer to as the "petroleum syndrome," namely an economic digestion crisis, characterized by excessive dependency upon oil for government revenue (over 50 percent) and for foreign exchange (over 75 percent). The revaluated bolivar raised the world market prices of Venezuela's traditional agricultural exports (coffee and cacao) to the point that they were no longer competitive. The lion's share of royalty income was pocketed by an irresponsible and corrupt regime. Secondary beneficiaries were the oil workers (3 percent of the labor force), the new oil bureaucrats, and the merchant satellites of the petroleum sector prosperity. A rapid exodus from rural to urban and petroleum areas compounded already serious housing, health, food production, and education problems—for none of which the Gómez government took responsibility.

By 1931 the world depression caused a 40 percent drop in crude prices and a 15 percent cutback in output. After 1928, the various producers were no longer rivals, for in that year Walter Teagle and Henri Deterding, heads of Standard of New Jersey and Shell respectively, had concluded a market-control agreement at Deterding's Achnacarry Castle in the Scottish highlands. Essentially, it was a prorationing pact that was soon adopted by the other big oil companies operating in third world areas. It was designed to maintain profits by avoiding destructive competition resulting from overproduction and price cutting. Thus was the international oil cartel born.[12]

One effect of the depression was to elevate Standard of New Jersey to the dominant position in Venezuela. Before that time it had invested over $20 million in dry holes. When the depression-beleaguered independent producers in the United States demanded protection against destructive foreign competition, Congress, on June 6, 1932, accommodated them with a high tariff on imported petroleum. As a result, Standard Oil of Indiana, having only United States market outlets, was forced to sell its Venezuelan holdings and its adjacent Aruba refinery to Standard of New Jersey, which became Venezuela's second-largest producer in 1932 and first by

1935, when its Creole subsidiary opened up the Oficina Fields in the eastern llanos.[13]

Though Venezuelan production returned to 1929 levels by 1934, the depression-induced crude price drop had sharply cut government revenues. As a consequence, Gómez brought Torres back into the Development Ministry to increase the government's share of oil industry profits. In August of 1930, he issued a *reglamento* that imposed state fiscal and technical controls over the producers. The following year he challenged the companies "high cost" and "low market price" figures and presented them with a huge bill for back taxes and royalties. Shell and Standard officials thereupon paid Gómez a visit at his headquarters in Maracay and came away praising his "statesmanship," for the *reglamento* was abolished and Torres was fired again.[14]

During the Gómez years, from 1922 to 1935, the companies extracted from Venezuelan soil 1,262 million barrels of oil, less than two years of current production. They marketed it for $1,670 million, of which the Venezuelan treasury received $118 million (7 percent), less than two days of current government income from the industry.[15]

Today's oil bonanza might not have been possible had it not been for the dictator's liberal inducements to the companies. His generosity brought them in to build up the industry rapidly, and once there with their big investments in production facilities they were captive and unable to resist persistent government demands for a greater share of their profits. No one, of course, condones the waste, the corruption, the irresponsibility, the lack of controls, and the failure to utilize the oil revenues for the benefit of the nation.

López and Medina: A Decade of Transition (1936–45)

When Gómez died, his successor, War Minister General Elias López Contreras, "inherited an oil factory," wrote Betancourt in his *Política y petróleo.*[16] He also inherited a poverty-stricken citizenry, an underground political opposition, and a long-frustrated coterie of nationalistic congressmen.

Nestor Luis Pérez emerged from one of Gómez' foul prisons to assume the post of development minister. Henceforth, he declared, there would be competitive bidding on all new concessions; the producers would be required to refine in Venezuela rather than in the adjacent Dutch islands; fiscal and technical controls would be placed upon the industry; the companies would be billed for excessive cost claims and for the lower-than-market prices declared during the Gómez era. The president, the congress, and the press unanimously supported Pérez. The companies stood firm and sought refuge and delay in the courts.[17]

President Lázaro Cárdenas' March 18, 1938, expropriation of Shell and Standard in Mexico put a scare into these same companies in Venezuela and gave hope and courage to the Venezuelan nationalists. The new July 13, 1938, petroleum law required competitive bidding on new concessions and set minimum royalties at 15 percent (instead of 10 percent or far less on Shell's holdings).[18] But it had no effect, for under such terms the companies applied for no new concessions. They were content to operate under the Gómez-era terms on existing concessions. Only in a token way was the government, under the 1938 law, able to increase its participation in the profits of the industry. It did this by extended fiscal and technical controls that reduced company cost claims and slightly raised "market prices" through an improved calculation method. Also, industry customs exemptions were cut back. The 1938 law's authorization for the government to enter the oil business came to nothing, however, for the capital requirements, the technical know-how, and the marketing outlets were all missing.[19] Unlike Mexico, Venezuela had to dispose of nearly all of her output abroad.

López Contreras' attempts to achieve adjustments for Venezuelan labor at the expense of the companies were similarly unsuccessful. His cautious permission of trade union activity led to the rapid formation of the Petroleum Workers Syndicate, a company refusal to deal with such an organization, a December 14, 1936, strike, a 40 percent drop in production, and consequently a 40 percent drop in government royalties. By January 22, 1937, the president had tolerated enough; government troops broke the strike, the syndicate was abolished, the labor leaders were exiled, and the workers were granted a mere twenty-two-cent increase in their daily wages.[20]

In 1938, López Contreras announced a grand three-year plan to "sow the petroleum," that is, to plow back the nation's oil income into economic and social development. Although production increased in 1937 and 1938 by about 20 percent, to 185 million barrels, market prices remained depressed. Demand in the principal

United States market rose following the Mexican expropriation, so production reached 206 million barrels in 1939. But then the adverse effect of World War II upon West European markets and upon tanker availability forced production back below 1939 levels until 1941.[21] Hence, surplus funds for "sowing the petroleum" were scarce. Besides, López Contreras' initial enthusiasm for social and political reform quickly waned when the popular forces thereby unleashed threatened the stability of his regime. Hence, López Contreras, in the Gómez tradition, ended his reign by continuing the despotism. In the words of Betancourt, in April of 1941, "López Contreras transferred the government of Venezuela to his own Minister of War, Gen. Isaías Medina Angarita. He thus continued fulfilling . . . the norms of an electoral system . . . whereby the presidency of the republic was the ultimate goal in a military career."[22]

As the United States prepared to enter World War II, it alleviated the tanker shortage and raised demands for Venezuelan crude. Production jumped 22 percent in 1941, from 186 million to 228 million barrels. But at year-end came Pearl Harbor and the subsequent movement of Nazi submarines to the Caribbean to cut off the Allies' Maracaibo Basin petroleum supply. Seven shallow draft tankers were torpedoed on the night of February 14, 1942, upon leaving Lake Maracaibo, and the Aruba and Curaçao refineries were shelled.[23] This, combined with the more urgent need of Shell and Standard's ocean tankers in the North Atlantic and Western Europe, resulted in another transportation shortage for Venezuelan oil. As the storage tanks filled, the producers had no alternative but to shut down wells. 1942 production dropped to 148 million barrels, 25 percent less than the previous year and the lowest output since 1934.[24]

The resulting one-fourth drop in government revenues prompted President Medina to try to obtain more income from the companies. His dutiful Congress quickly passed the 1942 income tax law, the first in Venezuela's history. Since it applied only to very large earners, just the oil companies were affected. The 2.5 percent maximum rate on net profits produced only $6 million in additional income— a mere fraction of the amount needed to alleviate the government deficit.[25]

Medina then approached the companies directly. He asked for "just participation," pointing out that Shell and Standard were still paying royalties of only about ten cents per barrel in Venezuela whereas in Mexico just before the 1938 expropriation they were paying thirty-eight cents. When the companies refused to help, he indicated to Shell that he would challenge its option to renew the thirty-year titles granted in 1912. He threatened Standard and Gulf with claims for back taxes for overstated costs and understated market prices. The level of intimidation was raised further by Medina's nationalistic speeches to enthusiastic Venezuelan audiences.[26]

The companies came to see the wisdom of compromise. Their lawyer, Herbert Hoover Jr., and Development Minister Eugenio Mendoza drafted a new law without any public or congressional participation. It was presented to the Congress with the admonition to pass it without changes lest the whole compromise be destroyed. Congress did so, under protest from a minority group of opposition deputies.[27]

The 1943 Oil Law, under which all concessionaires were henceforth required to operate, doubled royalties to 16⅔ percent, a figure calculated to assure the government one-half the net profits of the industry. In addition, new concessionaires were obligated to refine in Venezuela a minimum percentage of their domestic production. Also, the government was authorized to assume greater fiscal and technical control of the industry. The quid pro quo for the companies was new forth-year titles on all existing concessions and a government guarantee to drop all claims for back taxes.[28]

With a new law on the books, with transportation shortages easing, and with accelerated consumption of oil by the Allied war machine, the demand for Venezuelan petroleum grew. Production more than doubled between 1942 and 1945, the result of opening shut-down wells and drilling new ones. In 1944 and 1945 Medina had opened a new round of concessions bidding and granted nearly twelve million hectares, more than double the concessions area in existence in 1943. The government's oil revenues, as a result of the 1943 royalty increases and the proceeds from the concessions sales, were eight times greater in 1945 (614 million bolivars) than in 1942 (75 million bolivars).[29]

The resolution of its revenue problems was insufficient to save the Medina regime, however, for the continued despotism, the growing corruption, the neglect of social reform, the dissatisfaction with petroleum policy, and the noncirculation of the rising political and military elites resulted in growing public dissatisfaction with the government. On October 18, 1945, frustrated young army officers combined with leaders of the reform-minded, populist Acción Democrática (AD) political party and overthrew the Medina administration.[30]

Acción Democrática (October 1945–November 1948)

True to their word, the military participants in the new junta allowed AD politicians to dictate reform policies and administer the government. Party President Rómulo Betancourt, in conjunction with AD ministers, ruled by decree until February 15, 1948. Then AD President Rómulo Gallegos and the AD Congress ruled constitutionally until November 1948, when the young military officers seized the government for themselves.

In addition to its democratic political reforms, AD launched a comprehensive program for rapid development of the nation's human and physical resources. As this involved extraordinary expenses, a more aggressive petroleum policy became necessary. In charge of the Oil Ministry was Juan Pablo Pérez Alfonzo, a young professor of civil law at Central University in Caracas and an AD deputy in the Medina Congress. He had first gained notoriety for his opposition to the 1943 petroleum law. He assailed it in Congress for surrendering unnecessarily to the companies on the matter of back taxes and other pending claims. In addition, he warned that 16⅔ percent royalties were insufficient to guarantee the government one-half the profits because 1943 was a year of unusually low market prices.[31] When prices rose in 1944 and 1945, Pérez Alfonzo's warning was recalled.

Now, as oil minister, he moved immediately to guarantee the nation its 50 percent. The device was a December 31, 1945, presidential decree imposing extraordinary additional taxes on the companies up to the point that the government received one-half the industry's net income.[32] Similar decrees at year-end 1946 and again in late 1947 enabled the government to get its half, and when constitutional government was restored in 1948, the AD Congress put the 50–50 concept into law. Lest Venezuela's sharply rising taxes further increase its comparative cost disadvantage with the Middle East, its diplomats urged all other oil exporting nations to impose 50–50 taxes. Iran did so in 1949, Saudi Arabia in 1950, Kuwait in 1951, and Iraq in 1952, and by the mid-1950s all the world's oil exporting nations had adopted Venezuela's 50–50 tax initiative.[33]

Pérez Alfonzo saw an opportunity to raise the government's share by utilizing his option to take royalties in kind rather than in cash. On June 26, 1947, he announced that 25 percent of the 1948 and 1949 royalty oil (about 36 million barrels) would be auctioned in the open market.[34] While this device netted Venezuela an additional 3 percent of the industry's profits in 1948 when oil was in short supply, a buyer's market returned in 1949 and ended the practice.

Pérez Alfonzo also promoted the notion that the oil companies should assume responsibility for economic and social development. Under pressure from the nationalistic oil minister, Shell and Standard in 1947 began building 40,000- and 50,000-b/d refineries on the Paraguana Peninsula, invested $15 million in Nelson Rockefeller's International Basic Economy Corporation (which developed livestock, fishing, and supermarket cooperatives), and dropped domestic gasoline prices to the lowest in the world.[35]

The chief nongovernment beneficiaries of the AD junta's social reforms were the petroleum workers. Labor, the key political prop of the party, had been encouraged to organize and bargain for its rights. When the companies balked at the Petroleum Workers Syndicate's "excessive" demands, Labor Minister Raúl Leoni pressured them to sign the June 14, 1946, collective contract, which included substantial real wage increases and new fringe benefits for the workers. And when this expired at the end of 1947, the government again insisted that the companies sign a new three-year collective contract, which embodied additional gains for the workers.[36]

Pérez Alfonzo also enforced his conservation policies on the industry, warning that the companies could expect no more concessions. He condemned wasteful gas flaring and offset drilling and ordered both reduced.[37] He wanted to cut back production to preserve more income for future generations. However, the companies had an expanding market for Venezuelan oil in the post-World War II reconstruction of Western Europe, and President Betancourt and the AD Party had an expanding need for funds to finance ambitious programs of social reform and economic development. Accordingly, Venezuelan production rose another 50 percent (from 323 to 490 million barrels) between 1945 and 1948.[38]

The AD government came to an end with the military coup of November 24, 1948. Its aggressive petroleum policy did not precipitate its downfall. Rather, AD's overthrow was the result of its over-hasty reforms, its vindictive policies toward the old landed and political oligarchies, its neglect of rising new economic elites, and its political exclusivism. Its civilian opponents combined to encourage the military to take power.

The Pérez Jiménez Dictatorship (November 1948–January 1958)

Although exiled AD leaders claimed that the oil companies cooperated with the military in bringing about their downfall, no evidence to substantiate such charges has been uncovered.[39] There is no doubt, however, that the companies were relieved to have the radical populist AD government replaced by a military junta. The foreign oil companies' experience in underdeveloped countries throughout the world up to 1950 had clearly demonstrated that corporations prospered best under dictators. The taxes would be lower, labor would be disciplined, government technical and fiscal surveillance would be reduced, conservation would be ignored, and profits would rise. In none of these expectations were the companies disappointed by dictator Marcos Pérez Jiménez.

One of his first official acts was to dissolve the National Confederation of Workers and all its affiliates. This meant that the Petroleum Workers Syndicate existed no more, that the companies could return to the Gómez-era practice of refusing to recognize unions, and that all labor problems would be resolved by a ruler sympathetic to the needs of the companies.[40]

Early in 1949 the new Mining Minister Pedro Ignacio Aguerrevere announced, "The government may lighten the tax burden on the oil companies." In 1951, it negotiated a new reference pricing agreement favorable to the companies. The 50–50 law remained on the books but was not implemented fully. Betancourt calculated that whereas AD had raised the government's "take" to 53 percent by 1948, it declined to 44 percent under Pérez Jiménez by 1954.[41]

AD's aggressive nationalism toward the industry was now abandoned. Its 1948 plans for setting up a state oil company to exploit the national reserves was shelved. These parcels were now farmed out for development to the companies under liberal service contracts. No moves were made to increase domestic refining. Government fiscal and technical surveillance was sharply reduced. Gas flaring and offset drilling increased. Government auditing of company books stopped.

Finally, the government opened up Venezuela once more to new concessionaires. In rationalizing such a move, Pérez Jiménez explained on February 11, 1956: "We must prevent our oil investment capital from going to other countries. Our reserves must be maintained or increased."[42] The new concessions round began four months later. Competitive bidding produced $686 million for the government, but in the process about two million acres, in addition to the fourteen million acres under concession at the end of 1955, were granted to the companies for forty years.[43]

Given such encouragement, Venezuelan production rose rapidly under the dictator. In 1948, just before the coup, the oil companies, in reaction to the AD's aggressive nationalism, had cut production 15 percent (200,000 b/d), laid off 18,000 petroleum workers (about one-third of the total), cut off further investment in Venezuela, and shifted all new investments to the Middle East. But the abandonment of nationalism by Pérez Jiménez brought them back, and production more than doubled (from 1339 barrels in 1948 to 2779 million barrels in 1957).[44]

The growing despotism of the dictator, the accumulating evidence of massive fraud and peculation, the wasteful and irresponsible public spending, the surrender of petroleum nationalism, and the abandonment of responsibility for "sowing the petroleum" fostered the growth of massive public opposition and the alienation of the navy, which combined to topple the dictatorship in January 1958.[45]

For one year, 1958, a transitional junta headed by Admiral Wolfgang Larrazábal and supported by a coalition of civilian party leaders ruled Venezuela. Inasmuch as dismantling the dictatorship and restoring constitutional government were its assigned tasks, petroleum policy was not of primary concern. However, just before it terminated its work, on December 20, 1948, the junta raised the 50–50 tax law to 60–40 in the government's favor. This was done both to reduce the deficit and to erase the financial burden of its popularly elected successors, President Rómulo Betancourt and the AD Congress.[46]

Return of AD: A Decade of Controversy (1959–69)

When Betancourt returned to power as constitutional president (1959–64), he and his party quickly demonstrated that they had learned some fundamental lessons about political survival in Venezuela. Those included: compromise and cooperation with other political parties (this meant coalition government); sympathetic response to the salary, equipment, and force level demands of the military; recognition of the rural oligarchy's property rights (agrarian reform was acceptable only after paying the owner a fair market price for his land); respect for the vested interests of urban commercial and industrial elites (this meant minimizing their taxes and curbing labor demands). The radical populism of the 1945–48 era was abandoned. Economic development and social reform would now come at a more moderate pace. The costs of such programs would no longer be shared by the domestic private sector. Rather, they would be borne exclusively by the foreign oil companies.

AD's Mining Minister Pérez Alfonzo had also been sobered by his policy experiences of the 1945–48 period. During his decade in exile in the United States and Mexico, he devoted this time to diligent study and emerged an expert both on the Mexican nationalization experience and on the international oil cartel—particularly its internal functioning, its cost calculations, its production controls, and its marketing monopoly. Thus he returned to office a better strategist.[47]

To fund AD's economic development and social reform programs, petroleum policy was utilized to maximize income from the industry. The times, however, were far from propitious for raising the government's share. The oil industry resented the Larrazábal junta's last-minute 60–40 decree and threatened to disinvest in Venezuela and shift to the Middle East unless the Betancourt government rescinded it. Pérez Alfonzo countered this move by urging Middle East exporters to raise their taxes to the new Venezuelan level.[48]

New difficulties came in February 1959, when the oil cartel, faced with an oil glut from exploding Middle East production, dropped posted prices by eighteen cents per barrel, or about 10 percent. Since this reduced government revenues, Venezuela and all other oil-exporting nations protested, but to no avail. The glut increased in 1960 as independent producing companies began opening up Libya and entered the Middle East and Venezuela. Hence the Seven Sisters cartel began losing control over production. In response, Exxon announced in July 1960 that it would drop posted prices another ten cents per barrel. Pérez Alfonzo retaliated by giving preference to the independents on service contracts to produce from national reserve parcels. His anticompany public rage made relations so tense that both Standard and Shell decided to exempt Venezuela from their August 1960 price reduction.[49]

Yet another threat came in 1959 from import restrictions in Venezuela's principal market. The United States Trade Agreement Extension Act of 1955 had authorized the president to limit imports of those commodities that arrive "in such quantities as to threaten or impair national security." When the world petroleum glut loomed in 1958, the United States Independent Producers Association pressured the Dwight D. Eisenhower administration to impose import restrictions on foreign oil. In May 1959, the United States established a quota licensing system that limited imports to 13 percent of domestic consumption but exempted overland imports from Canada and Mexico. The Betancourt government charged discrimination and pleaded for fair treatment. It also argued that the preference of Venezuelan over Middle East oil was vital to hemispheric security. But the companies, which made higher profits on Middle East imports, convinced Eisenhower to reject such preference. Hence, Venezuelan oil remained restricted in the United States market until the quota system was abandoned during the 1973 energy crisis.[50]

Inasmuch as the world oil glut and marketing restrictions had removed any leverage Venezuela might have to increase its percentage tax share, Pérez Alfonzo was forced to adopt alternative strategies to increase income. Early in 1960, he established the state Corporación Venezolana de Petroleo (CVP), which obtained, as a result of competitive bidding, over 80 percent of the net profits on oil produced from national reserve concessions under long-term service contracts.[51] Ever suspicious that Venezuela was being cheated by devious company accounting, he established a Coordinating Commission for Conservation and Commerce of Hydrocarbons to regulate company production and to audit their pricing and marketing methods. When the commission revealed some abnormal discounts on sales of Venezuelan oil, Pérez Alfonzo ordered production suspended on any concession from which oil was henceforth sold at less than market prices.[52]

Conservation, however, was the commission's main function and the principal long-range concern of Pérez Alfonzo. It was made unmistakably clear to the companies that the government would never grant more concessions. Commission technicians went into the major fields to monitor production techniques. They ordered reduced flaring and increased utilization of natural gas. They curbed offsetting and regulated flow to prolong well life and to get the maximum recovery from the fields.[53]

The reaction of the companies to such militant pricing, marketing, and production control policies was to cut back their operations in Venezuela and to increase them in the friendlier Middle East. Drilling declined, employment dropped, and annual reinvestment rates declined by 28 percent under the Betancourt presidency. Production increased at an average rate of only 3 percent, from 2.7 million b/d in 1959 to 3.2 million in 1963.[54]

Although Venezuelan businessmen criticized these nationalistic government policies, which also hurt their businesses as the oil industry declined, Betancourt and Pérez Alfonzo were unmoved. When they adopted their basic conservationist-nationalist policies of increasing the government's income from existing production, they fully anticipated that the companies would shift their operations elsewhere. They deliberately provoked this, for they were determined to reduce the nation's extreme dependence upon oil. In the judgment of an astute policy analyst, Peter R. Odell,

The rationale for this policy stems from the belief that the oil sector of its economy is too dominant—accounting for 90 per cent of its exports and about 20 per cent of the G.N.P.; but at the same time giving employment to only two per cent of the labour force in a situation in which a population growing at a rate of about 3.5 per cent per annum demands a large number of new job opportunities. The oil industry, moreover, the Government argues, is in foreign hands with no degree of Venezuelan control over decisions taken in New York or London but which vitally affect the Venezuelan economy.[55]

The most fundamental of Pérez Alfonzo's policy reforms, however, and the one which ultimately led to nationalization in 1976, had little or no impact while AD was in power during the 1960s. This was the founding of OPEC. This move was triggered by the oil cartel's February 1959 10 percent reduction in the posted prices. This prompted Saudi Arabian Oil Minister Abdullah Tariki to call for a meeting with his oil-exporting neighbors—Iraq, Kuwait, and Bahrain. Venezuela was invited to send observers. Pérez Alfonzo arrived at the April 1959 First Arab Oil Congress armed with interpreters and Arab translations of Venezuelan laws and policies. Here he began to advocate the strategy he had learned in exile from the Texas Railroad Commission, which had since the 1930s regulated output by prorationing agreements among the independent oil producers, and in that manner kept prices from falling in the United States. Pérez Alfonzo now advocated international prorationing. The way for the oil-exporting countries to get more was to sell less, he declared. This was just the opposite of the company strategy, which was to sell more by lowering prices to increase consumption. Having convinced the Arabs, Pérez Alfonzo flew to Tehran and convinced the Shah. When in August of 1960 the oil cartel announced yet another drop in posted prices, the oil ministers of Saudi Arabia, Iran, Venezuela, Iraq, and Kuwait assembled at Baghdad and founded OPEC.[56] "We can no longer remain indifferent . . . to . . . price modifications," they declared. "We demand that the oil companies return present prices to the levels prevailing before the reductions . . . and maintain their prices steady." At the same time the ministers agreed to set up a permanent organization, which other oil exporting nations were invited to join, "to formulate a system to ensure the stabilization of prices by, among other means, the regulation of production."[57]

Pérez Alfonso was the most aggressive promoter of OPEC for very good reasons. The world oil glut and subsequent price wars had placed Venezuela at a cost disadvantage with the Middle East. In addition, the latter's 70 percent share of proven world reserves compared with Venezuela's 7 percent share explains Pérez Alfonzo's special concern for conservation. But despite his persistent efforts, OPEC failed to work for a full decade. Though world oil consumption skyrocketed during the 1960s, production increased even faster as independent producers broke the monopoly of the Seven Sisters cartel, and new nations, particularly in Africa, became big exporters. Attempts at price and production controls in the Middle East area were defeated by Arab rivalry and factionalism and by the impatience of Arab and Iranian rulers to gain maximum incomes quickly. In 1962, the Saudi king replaced price and conservationist hawk Tariki with the more moderate Zaki Yamani, and in late 1963, as President Betancourt neared the end of his term, Pérez Alfonzo himself resigned. Up to that time OPEC had achieved none of its goals. The companies made their 1959–60 price reductions stick, and posted prices remained around $1.80 per barrel until the end of the decade.[58]

AD retained control of the government during the late 1960s, as

Betancourt elevated his Labor Minister Raúl Leoni to the presidency (1964–69) and Pérez Alfonzo placed his protégé Manuel Pérez Guerrero in charge of the Mining Ministry. Hence there was little policy change. The reforms and initiatives of Pérez Alfonzo on conservation, taxation, posted prices, and government controls all continued. Despite his "retirement," Pérez Alfonzo continued as the dominant policy-making influence. Just as before, though the companies resisted AD nationalism by expanding production everywhere except in Venezuela, where it remained steady at about 3.5 million barrels daily, the Leoni government continued to increase Venezuela's "take" from existing production. In 1965, Pérez Guerrero's Coordinating Commission for Conservation and Commerce of Hydrocarbons intensified its audits of the companies, again found them guilty of marketing Venezuelan oil at artificially low prices, and claimed $110 million for back taxes. When they resisted payment, Pérez Guerrero threatened to amend the tax law to include Pérez Alfonzo's long-held and well-publicized proposition that anything over a 15 percent return on company investment must be returned to the nation as excess profits. He also threatened to extend CVP functions to compete with the companies in production and domestic marketing. In the face of such pressures the companies saw the wisdom of compromise. In July 1966 they agreed to pay retroactive tax claims in the amount of $155 million, to base future taxes on five-year fixed reference prices (thus resolving the discount sales controversy), and to increase production by at least 3 percent annually. In return, the government abandoned its threats to impose an excess profits tax and to enter the oil business itself. The continued glut in the marketplace, however, restricted the government's increase in the taxable profits of the industry to just 3 percent (from 65 percent up to 68 percent) under the Leoni administration.[59]

Toward Nationalization, 1970–76

The oil companies breathed a sigh of relief when the centrist Christian Democrats (COPEI) won the December 1968 elections and Rafael Caldera assumed the presidency in 1969. Caldera had backed off during the campaign from AD's aggressive nationalism. He had argued that the petroleum investment decline and production stagnation had injured domestic business and labor interests, and he indicated little interest in furthering state surveillance and control over the industry.

Such relief was short-lived, however, as AD emerged as the dominant force in Congress, and Pérez Alfonzo's influence remained preeminent via the AD congressional leadership of Arturo Hernández Grisanti. While Caldera hesitated and vacilated, Congress wrested control of petroleum policy from him. When, for example, Caldera attempted to increase production by offering the companies liberal service contracts on national reserve parcels in 1969, Pérez Alfonzo publicly objected, and Hernández Grisanti pushed through Congress a bill requiring up to 85 percent profit for the nation on such contracts. Under these terms, the companies lost interest and production did not rise. This was just the outcome conservationist Pérez Alfonzo desired. The following year when Caldera tried to augment the government's revenues by new sales taxes on domestic business, Pérez Alfonzo and Grisanti blocked this move by insisting that any increased taxes be shouldered solely by the oil companies. And again, Pérez Alfonzo insisted that company profits should not be permitted to exceed 15 percent.[60]

In the midst of these domestic policy disputes, Venezuela was overtaken by an international event so worldwide in its repercussions that it put the extreme nationalists in the driver's seat and made their ultimate triumph inevitable and complete. This was Libyan dictator Col. Muammar el-Qadaffi's August 1970 shattering of the oil cartel's $1.80 per barrel price front. What had happened was that world oil consumption, which in the post-World War II cheap energy era increased from 10 million barrels daily in 1950 to 60 million by 1970, finally caught up with production. When the half-century-long buyer's market suddenly became the seller's, Qadaffi put pressure for price increases upon the most vulnerable link in the industry's price holding chain, namely Armand Hammer's Occidental Petroleum Company, which was wholly dependent upon Libyan oil to supply its West European market. When Hammer capitulated and agreed to pay $2.10, Qadaffi, threatening nationalization, forced his increase on Standard of New Jersey, Shell, Standard of California, and Texaco. With the cartel price control broken, the leapfrogging began as Iraq, Algeria, Iran, and Kuwait raised their prices to match Libya's. "Fifty-fifty" arrangements jumped to "55-45" in the Middle East in the fall of 1970.[61]

At the OPEC Ministers meeting in Caracas in December 1970, Pérez Alfonzo insisted that President Caldera impose Libyan prices immediately, in violation of President Leoni's 1967 agreement for five-year fixed reference prices. The OPEC ministers unanimously adopted 55 percent minimum government profit shares and ratified the new Libyan price levels. The following month, in Tehran, they approved the Shah of Iran's demands for fifty-four cents per barrel more. In March 1971, the Venezuelan Congress raised reference prices by fifty-nine cents. In April, Qadaffi signed with the fifteen oil companies operating in Libya a five-year agreement to fix posted prices at $3.45 per barrel. This so-called Tripoli agreement was quickly extended to the twelve other members of OPEC.[62]

These developments merely confirmed the shift in production, price, and market control from the companies to the exporting countries, all of which, as a result, began to anticipate early nationalization. Pérez Alfonzo first raised the issue in the spring of 1971 and urged speedy action by Congress to protect the national interest. He warned of the danger that the companies might deplete their fields and let their plants and equipment run down as their concessions approached expiration in 1983. To guard against this, Congress on July 30, 1971, passed the Hydrocarbon Reversions Law, which required the companies to post bonds to guarantee the turning over of their facilities and fields to the government in good condition. The following month Venezuela nationalized natural gas.[63]

The companies' retaliatory response to this persistent nationalization was to cut back Venezuelan production 15 percent (from 3.8 million b/d in 1971 to 3.2 million by 1973). Caldera threatened them with minimum export volume penalties. Pérez Alfonzo argued against this, however, for he wanted production to drop for conservation reasons, and he urged reference price increases instead.[64] Again, international events overwhelmed and resolved the Venezuelan domestic policy debate. This time it was the October 1973 Arab-Israeli War, which, because of the ensuing Middle East oil embargo, broke all company resistance to increasing production in Venezuela. It also quadrupled world oil prices, to $14 per barrel; this ended Caldera's revenue concerns as Venezuela's tax take rose to $10 billion in 1974 (from $1.5 billion in 1970).[65]

When newly elected AD President Carlos Andrés Pérez took office early in 1974, he responded to his party's impatience in Congress by calling for nationalization of the oil industry at the earliest possible moment. Most concessions still had nine years to run, but the companies did not protest. For them, Venezuela was becoming a less profitable place to do business than the Middle East, and they showed this by reducing output by a third between 1973 and 1975 (from 3.3 million to 2.2 million b/d). Not only were Venezuelan producing costs greater, but its taxes were higher (they rose from 65 percent of profits in 1972 to nearly 90 percent by 1975), its labor unions were more troublesome, and its government was more nationalistic.[66] Besides, the companies expected to play a role in the production and marketing of Venezuelan oil in the postnationalization era that would allow them an income, because of high market prices, equal to that when they owned the oil themselves.

In the summer of 1975 when President Pérez submitted the nationalization bill to Congress, however, the more ardent nationalists vehemently opposed Article 5, which allowed the state company, Petróleos de Venezuela (Petroven, soon renamed PDVSA), to grant the oil companies production, sales, and technical assistance contracts. The nation's number-one oil expert, Pérez Alfonzo, helped the president prevail by demonstrating that Venezuela was not yet ready to go it alone.[67]

The nationalization bill was passed by Congress on August 21, 1975, and was signed by the president that same day. All private concessions were to be terminated December 31, 1975, at which time all exploration and exploitation of petroleum would come under the exclusive control of the state. The latter's responsibilities were to obtain maximum economic yields for the nation. PDVSA would be converted into a mercantile company; its employees would no longer be considered public. It was allowed to produce and market either on its own or via service contracts with foreign or domestic companies. Though established as an autonomous agency, PDVSA was to remain under supervisory control of the Ministry of Mines. Under Article 12, the president was authorized to purchase the foreign company installations at their book value, which at one billion dollars was less than 20 percent of market value, but not to pay them anything for the profits they might have obtained had their concessions been allowed to run to 1983.[68] The companies threatened legal action over the terms of Article 12, but President Pérez deterred them by suggesting that such action might make them ineligible for service contracts and future rights to produce and buy Venezuela's oil.[69]

And so on January 1, 1976, Venezuela nationalized its oil industry. In contrast to the experience of Mexico in the 1930s, to Iran in the 1950s, and to Peru in the 1970s, this event was relatively peaceful and devoid of significant conflict. The act of nationalization was anticlimactic, for state control, regulation, and profit participation had increased so steadily under Pérez Alfonzo's aegis for the previous fifteen years that an ultimate government takeover was fully anticipated. OPEC had put the state in control by 1970, and most concessions were coming to an end in seven years anyway. Company income from the government-owned industry would continue at or near existing levels through the profits from the service contracts, oil-purchase agreements, transport arrangements, and sales to foreign consumers. Besides, the companies were now relieved of the long bitter adversary relationship with the Mining Ministry, the chronic discord over taxes, and the difficulties of coping with the militant petroleum workers. And finally, nationalism of all third world natural resources seemed by 1975 to be an unstoppable trend.[70]

Pérez Alfonzo: An Appraisal and Interpretation

Examination of the record leaves little doubt that Juan Pablo Pérez Alfonzo, for over thirty years (1945–76), was *the* architect of Venezuelan petroleum policy, even though he was oil minister for only seven of those years (1945–48 and 1959–63). He was responsible for nearly all the policy initiatives—taxes, prices, conservation, regulation, controls, and ultimate nationalization. And in the face of strong foreign and domestic criticism, his views generally prevailed. He obtained for his nation a greater share of industry profits than any other oil-exporting country in the world.[71] Through auditing procedures, fiscal controls, and new pricing mechanisms (substituting posted prices, and later reference prices, for actual market prices) he steadily escalated, ratchet-like, the government's take. His resistance, in disdain of market mechanisms, sheltered Venezuela alone from the price drop of 1960. He spread Venezuelan tax initiatives—50–50 in 1945 and 60–40 in 1959—to all other oil-exporting countries. And his conviction that the intrinsic value of petroleum would some day be recognized was ultimately realized through OPEC, which he fathered and persistently promoted. Although Venezuela had long been regarded by Latin American nationalist writers as a classic client regime of foreign oil companies, the fact is that Pérez Alfonzo began to change all this long ago. From 1945 onward, he steadily forged a more and more advantageous relationship for his country in relation to the multinationals.

Obviously, Pérez Alfonzo would have been able to achieve little or nothing without the confidence and backing of Venezuela's dominant politicians, principally Betancourt, and its main reform party, AD. In his excellent and perceptive *Politics of Oil in Venezuela*, Franklin Tugwell concludes that Pérez Alfonzo's policy success was "due to a remarkable mixture of acumen, skill, and outright luck" and that by a strategy of assertive experimentation combined with national flexibility, Venezuela was able to capture an ever-increasing share of industry profits "without significantly hampering the production process."[72] Though Tugwell also attributed this success to fortuitous changes in the international market situation (who could predict Qadaffi's price-shattering coup of 1970?) Pérez Alfonzo created a good deal of his own luck by his international policy initiatives.

In large measure he succeeded because of his expertise. His pre-World War II career in civil law stood him in good stead for adopting sound legal stratagems. His frustrating experiences, stemming partly from inadequate knowledge of the oil industry during his first ministry, were overcome by a decade of intensive study while in exile in the United States and Mexico. Although when he returned as minister in 1959, he knew more about oil than any other Venezuelan, he employed the best international consultants—particularly Walter Levy and the top specialists from the Texas Railroad Commission—to advise him.[73] He was no great innovator. For nearly all his policy undertakings he found precedent elsewhere. Pemex, established in 1934, was his model for the CVP, and for his technical and fiscal controls and for taking royalties in kind he emulated The Texas Railroad Commission and the United States Independent Oil Producers Association with their prorationing and pricing strategies.[74]

Not so generally appreciated and understood, however, were Pérez Alfonzo's extraordinary political skills. Quite deliberately he provoked an adversary relationship with the foreign oil companies. His perjorative rhetoric concerning the companies' selfish "plundering" of the Venezuelan nation, their "dishonest" business practices, and their "obscene" profits began with his opposition to the

1943 oil law and became ever more shrill and vituperative during the 1950s, 1960s, and 1970s. As a nationalist propagandist, he had no equal. As minister he made full use of the media—the weekly press conference, radio talks, and television—to belabor the industry. As a private citizen, he not only continued to use the media but also issued a steady stream of polemical books—an even dozen up to the time of nationalization[75]—and he worked through his protégés in the Mining Ministry and through his AD colleagues in Congress to propagate incessantly his nationalistic oil dialectic.[76] His critics, both foreign and domestic, labeled him a demagogue. No matter, for Venezuelan opinion had been won overwhelmingly to his side. His deft touching of the raw nerve of nationalism in Congress is what ensured his policy triumphs.

The wellspring of his overall policy was a universal ecological conscience. Oil to him was a vital, precious nonrenewable resource. Hence, his castigation of waste, his insistence upon reducing production and consumption, and his admonition to all governments to preserve it for future generations. An austere moralist, he was ever quick to condemn both the oil companies and the politicians for squandering the oil bonanza. His life-long conservation drive produced results: one-third of production was shut in by 1976, and the government had moved to alleviate the economic indigestion caused by the post-1973 avalanche of petroleum dollars by setting up the Venezuelan Foreign Aid and Investment Fund to isolate foreign dollars from the domestic income stream. But he was never satisfied. What had been done to conserve and to stop waste was not nearly enough. To the day of his death in September 1979, he lamented that the OPEC child he had fathered had turned into a prodigal son.[77] Belatedly, he realized that the Venezuelan tail could no longer sustain its role of wagging the Arab dog.

Postnationalization: 1976–83

PDVSA, the new national oil firm, became overnight, on January 1, 1976, the world's ninth-largest oil company and the sixteenth-largest industrial company in the world. Export sales in 1976 were $10 billion, making Venezuela the world's third-largest exporter, after Saudi Arabia and Iran. PDVSA had twenty-two thousand employees, five million acres of concessions, over $5 billion worth of plant and equipment (including twelve refineries and fourteen tankers), a production capacity of 3.3 million b/d, and twenty years of proven reserves at the 1976 production rate (2.2 million b/d). Its operating expenses, including $550 million annually for reinvestment, were but 15 percent of sales, the 85 percent remainder ($8.5 billion) accounting for three-fourths of the government's total revenues.[78]

PDVSA is a corporation owned by the Venezuelan nation. Its directorate includes a president, a vice-president, and nine directors, all of whom were initially selected on the basis of their managerial skills and technical expertise. It was designed as an apolitical body that reported directly to a presidential board headed by the minister of energy. Gen. Rafael Alfonzo Ravard, a model administrator and technocrat, was selected as PDVSA's first president. Previously he had achieved outstanding success as president of the Guayana Corporation, the big national industrial complex in southeast Venezuela. Initially PDVSA had fourteen subsidiaries, each encompassing the former operations of the various foreign oil companies. By 1979 these had been reduced to four: LAGOVEN (formerly Standard of New Jersey), MARAVEN (Shell), MENEVEN (Gulf), and CORPOVEN (Mobil and CVP).[79]

PDVSA had four main tasks. First of all, it had to assure the uninterrupted operation and smooth functioning of the industry, for the government had long been dependent upon a steady flow of oil revenues. Secondly, it had to modernize its plant and equipment, since the foreign oil companies, in the face of the growing nationalism since 1958 and in anticipation of reduced production and a gradual drawdown of their Venezuelan operations, had made minimal reinvestments. Hence, much of the installations were worn out or obsolete and needed replacement. Thirdly, it had to generate new reserves, for the companies had curtailed exploration since the 1960s, and as a result proven reserves had dropped to a low of only 20 years at the 1976 production rate. Finally, it had to develop Venezuelan oil expertise if the nation ever hoped to reduce its dependency upon the foreign technicians. To achieve these tasks, the 1976 nationalization act had provided a development fund under which 10 percent of export sales revenue was set aside for exclusive PDVSA use and control.[80]

Under the able presidency of Alfonzo Ravard, PDVSA ran smoothly from 1976 to 1979 and accomplished most of its goals. Careful long-term, collective-contract negotiations with the Petroleum Workers Syndicate avoided the disruptive labor troubles that had plagued Pemex in the aftermath of the 1938 nationalization in Mexico. Production and sales contracts for 85 percent of its oil were signed with Shell and Exxon. Their managers and technicians remained to produce for their companies about twenty-five cents per barrel in profits, about the same as in 1975. These arrangements were revised in 1980, so that the multinationals by 1983 accounted for only half of production and sales. This reflects both the rapid increase in newly trained Venezuelan technicians and the emergence of PDVSA as an independent marketer.[81] Relations between PDVSA and the Pérez administration from 1976 to 1979 were cordial, cooperative, and proper. Energy Minister Valentín Hernández respected PDVSA's autonomy. To the governing board of PDVSA, President Pérez appointed nine directors selected on the basis of their managerial ability, oil industry experience, and technical expertise.

A three-year honeymoon period opened the relations between PDVSA and the government. PDVSA did not remain immune from public criticism, however, especially from leftist politicians and public criticism, however, especially from leftist politicians and particularly from Pérez Alfonzo. In October 1978 he charged (inaccurately) that 1977 oil income was no greater than in 1973, that production was declining because of technical deficiencies (the main reason was declining reserves), that Saudi oil prices were higher than Venezuela's (the latter's was heavier crude and therefore worth less), that PDVSA was investing too extravagantly (he wanted to curtail rather than increase production), and that the multinational companies still controlled the industry (not true).[82] In general, however, the PDVSA operation received strong support from the Venezuelan people, political parties, and business community.

Unhappily, the same support was not given to the Pérez administration (1974–79) with respect to managing the petrodollar boom. Carlos Andres Pérez, the AD candidate, and his party had posted a landslide victory in the December 1973 elections, and when he took office to begin his five-year term in March 1974 the outlook seemed rosy. The year when OPEC first became an effective cartel, 1974, produced a windfall of $10 billion petrodollars. The government's income rose from 16 billion bolivars in 1973 to 43 billion in 1974. During the Pérez administration, the state received $229 billion in oil income versus $148 billion for the entire 1917–73 period (figures in 1973 constant dollars), that is, over 50 percent more than all previous Venezuelan governments combined since oil production began. "We are going to change the world," Pérez proclaimed as he inaugurated the most ambitious development program in Venezuela's history. The goal: to transform Venezuela quickly into a modern industrialized nation. The assumption was that now that the necessary capital was available, the "takeoff" towards industrial maturity, agricultural modernization, and prosperity for all was inevitable. Thus in 1974 Pérez launched his $54 billion five-year program. The state invested in huge public works projects, created new industries, and greatly expanded social welfare programs. By the end of his term Pérez proudly announced that his goals had been achieved.[83]

There were dissenting views, however. Already in 1976 Pérez Alfonso had these misgivings about the flood of petrodollars: "What is oil doing to us? . . . We are dying of indigestion. . . . Look at this waste and corruption. . . . We are putting our grandchildren into debt. . . . I call this 'Gran Venezuela Plan' the 'plan of destruction.' . . . We are drowning in the devil's excrement."[84] Which Pérez was right? Was the oil dollar boom a curse or a blessing?

The year-end 1978 results favored the latter view. After all the oil revenues, and more, had been spent, the nation appeared to have suffered a development disaster. Although real annual GNP rose 6.4 percent and manufacturing by 8.5 percent annually during these five years, the administration's huge public works, industrial development, and agricultural modernization projects overwhelmed and crowded out the private sector. Also, the results were unimpressive because of poor planning, waste, costly expenditures, low productivity, financial chaos, and corruption. Other results were rapid inflation, adverse trade balances, mounting debt, and increased oil dependency. The latter came as a shock in 1977 as the 15 percent projected annual increases in oil prices did not materialize and, despite PDVSA's best efforts, oil production fell from 3 million b/d in 1974 to 2.2 million in 1977. Although oil revenues were no longer sufficient to cover rising government expenditures, the Pérez administration proceeded to live beyond its means by going heavily into debt.[85]

The social results were equally bad. Real wages did not rise, and social inequalities deepened. Though government expenditures tripled under Pérez, the main beneficiaries appear to have been a bloated bureaucracy, for social welfare and social reform progress was not significant.[86]

But perhaps the Pérez administration's most serious failure was political, and for this the electorate avenged themselves in December 1978 as AD was turned out of office by the COPEI Party. The petrodollars had been utilized by Pérez to overdevelop the executive power of the state, distorting and interrupting a generation of democratic progress and political development. Overweening presidential power accumulation via oil money split the AD Party, alienated congress, weakened the party system, and disaffected the voters. This, added to the corruption scandals, the fattened bureaucracy, and the shortcomings in economic development and social reform, produced the electoral rejection.[87]

COPEI president Luis Herrera Campins assumed office in March 1979; as for Pérez five years earlier, circumstances augured well for his success. He too was the recipient of a petrodollar windfall stemming from a second oil price shock that followed the late 1978 Iranian revolution that deposed the Shah and removed five million b/d from the world's oil supply. The result was that oil prices more than doubled, to $36 per barrel, during 1979 and the government's oil income, despite only a 5 percent increase in exports, rose from $37 billion bolivars in 1978 to 58 billion in 1979, a 60 percent increase. Herrera rapidly increased expenditures to absorb the new income, now giving priority to social welfare rather than economic development. However, the new oil bonanza lasted little more than a year, for by 1980 the consuming nations, in reaction to the price rise, began to reduce consumption. Noncommunist world demand dropped from 51 million barrels per day in 1979 to 46 million in 1981. Also, by the latter year, Alaskan, Mexican, and North Sea production began to come on line and OPEC's share of the noncommunist world market fell 27 percent. OPEC supplied 60 percent of the free world's oil in 1974 but only 40 percent by 1981. Venezuela's export quota, set by OPEC, was reduced by 25 percent, from 2 million b/d in 1979 to 1.5 million b/d in 1982. Additionally, the oil glut weakened prices by 20 percent, from $36 per barrel in 1979 to $29 in 1982.[88] Venezuela, with heavy crude, was forced to sell for as much as $4 less, since its traditional United States market was sharply reduced by conversion to coal and by invasion of this market by price-cutting Pemex.[89]

The price drop and export decline reduced the government's oil revenues from $15 billion in 1981 to an estimated $10 billion for 1983. Herrera's response was to keep expenditures in place, deplete reserves from $20 billion to $9 billion, and increase the public debt by $10 billion, to $25 billion in 1983. The administration's reluctance to reduce mass consumption subsidies (especially on gasoline and food) or to cut back government expenditures in the face of the December 1983 elections appeared to be leading the nation toward financial disaster.[90]

Equally dangerous was the Herrera administration's political intervention in the PDVSA operation. The entering wedge was the August 1979 amendment to the nationalization law that required Energy Ministry approval of PDVSA budgets and reduced PDVSA directors' terms from four years to two. This was seen as a political power play by the new Energy Minister Humberto Calderón Berti and by Congress, for both began to criticize PDVSA's cost and growth plans, particularly its long-range heavy oil Orinoco development project, the so-called FAJA.[91] During 1980 Calderón Berti tightened Energy Ministry surveillance over the industry. In August 1981 he began making political appointments to the PDVSA board and started intervening personally in its meetings. Meanwhile congressional criticism of PDVSA expenditures, particularly high employee salaries, escalated. Both explained their attacks as a "deepening of nationalization," but the effect was to lower morale, as PDVSA directors and managers began to view politics rather than merit as the preferred route to high positions in the industry.[92] As the administration's financial problems mounted, Herrera began to eye the $8 billion PDVSA development fund. On September 28, 1982, he ordered the Central Bank to take it over, a clear violation of the 1976 nationalization act, for PDVSA now lost jurisdiction over its investments. The Orinoco project was frozen and $2 billion of the fund was used to bail out the corrupt Workers Bank.[93] Thus was the autonomy of PDVSA further reduced.

The Outlook

Venezuela's energy and political future will hinge on a number of variables. The government's income will remain heavily dependent on the vagaries of the international oil market. Until prices rise, its financial problems will continue.

It also faces an indefinite period of reduced production, until the

FAJA is developed. The main requirements for success here are that world consumption and market prices rise and that expansion costs do not become prohibitive. Obviously PDVSA's operation would improve if pre-1979 conditions were restored, that is, the return of its development fund, freedom from political interference, restitution of the merit system, and replacement of the political directors by better-qualified administrators.

On August 31, 1983, President Herrera appointed Energy Minister Calderón Berti to succeed Alfonzo Ravard as head of PDVSA for the period 1983–85. Now that AD has won the December 1983 presidential elections, PDVSA and the government will be run by rival political parties, a situation which augurs ill for the depoliticization of the national oil industry.

But most important of all for the future will be wiser administration, governments whose priorities combine rational, economic growth with steady social progress, and balanced democratic political development. After the multiple failures of the past decade, Venezuela stands in need of a political savior. Should the democratic parties continue to fail in their mission, there remains the danger of a return to military dictatorship.

Notes

1. Luis Vallenilla, *Oil: The Making of a New Economic Order: Venezuelan Oil and OPEC* (New York, 1975), pp. 73–76.

2. Anthony Sampson, *The Seven Sisters: The Great Oil Companies and the World They Made* (New York, 1975), pp. 63–70.

3. U.S., Bureau of Mines, *World Oil: Annual Report, 1920–1929* (Washington, D.C.).

4. Sampson, pp. 167–85.

5. Edwin Lieuwen, *Petroleum in Venezuela: A History* (Berkeley, 1955), pp. 10–15; B. S. McBeth, *Juan Vicente Gómez and the Oil Companies in Venezuela, 1908–1935* (New York, 1983), pp. 15–19.

6. Rómulo Betancourt, *Venezuela: Política y petróleo* (Mexico City, 1956), p. 43.

7. *Gaceta Oficial* (Caracas), July 19, 1922, p. 1.

8. Rómulo Betancourt, *Venezuela's Oil*, trans. Donald Peck (London, 1978), pp. 18–19.

9. Lieuwen, *Petroleum in Venezuela*, pp. 30–34; McBeth, *Juan Vicente Gómez*, pp. 76, 85.

10. New York *Times*, March 18, 1923, p. 13.

11. Lieuwen, *Petroleum in Venezuela*, pp. 38–49.

222   *Edwin Lieuwen*

12. John M. Blair, *The Control of Oil* (New York, 1976), pp. 54–56.

13. Lieuwen, *Petroleum in Venezuela*, pp. 56–60. For similar reasons Standard of Indiana was forced to bail out of its Mexican operation by selling its Huasteca subsidiary to Standard of New Jersey.

14. Ibid., pp. 65–67; McBeth, *Juan Vicente Gómez*, pp. 191–92.

15. Venezuela, Ministerio de Agricultura y Cria, *Memoria al congreso* (Caracas, 1936).

16. Betancourt, *Politicia y petróleo*, pp. 79–83.

17. Lieuwen, *Petroleum in Venezuela*, pp. 72–77.

18. *Gaceta Oficial*, July 14, 1938, p. 1.

19. Lieuwen, *Petroleum in Venezuela*, pp. 78–80.

20. Ibid., pp. 80–82.

21. *World Oil: Annual Reports, 1934–1941*.

22. Betancourt, *Política y petróleo*, pp. 153–63.

23. Stanton Hope, *Tanker Fleet: The Story of Shell Tankers and the Men Who Manned Them* (London, 1948), pp. 57–58.

24. *World Oil: Annual Reports, 1934–1941*.

25. Vallenilla, *Making of a New Economic Order*, pp. 46–47. Ultimately, not royalties but income taxes became the main device for increasing the government's share of the profits of the industry.

26. Lieuwen, *Petroleum in Venezuela*, pp. 92–95.

27. Betancourt, *Política y petróleo*, pp. 142–49. The law was passed on March 13, 1943.

28. *Gaceta Oficial*, March 14, 1943.

29. Venezuela, Ministerio de Hacienda, *Cuenta*, 1943–46 (Caracas, 1944–47). From 1945 onward the oil industry accounted for more than half of total government revenues.

30. See *El Pais* (Caracas), Oct. 18–31, 1945, for an analysis of events leading to the coup.

31. Venezuela, Congreso, Cámara de Diputados, *Diario de debates*, Feb. 23, 1943, p. 8, and March 5–6, 1943, pp. 4–6. Pérez Alfonzo's actual title was minister of development. A separate Mining Ministry, which included petroleum, was not established until the 1950s.

32. Venezuela, Ministerio de Hacienda, Administración General del Impuesto sobre la Renta, *Informe* (Caracas, 1946), pp. 35–36.

33. The British refusal to accept Iran's 50–50 demands precipitated the Anglo-Iranian crisis of the early 1950s.

34. *Gaceta Oficial*, June 26, 1947, p. 1.

35. Betancourt, *Política y petróleo*, pp. 264–76.

36. Because of the 50–50 arrangement, the government was actually paying half the new benefits received by the workers.

37. Betancourt, *Política y petróleo*, pp. 253–57.

38. *World Oil, Annual Reports, 1946–1949*.

39. For example, exiled President Rómulo Gallegos declared from Havana that responsibility for the November 1948 revolution belonged to "foreign capital and the petroleum interests" *El Heraldo* (Caracas), Dec. 12, 1948, p. 1.

40. Betancourt, *Política y petróleo*, pp. 523–24, 647–54.

41. Ibid., pp. 523–25, 678–93; Franklin Tugwell, *The Politics of Oil in Venezuela* (Stanford, 1975), pp. 47–48. Tugwell estimates that company return on investment rose from 14 percent in 1949 to 30 percent by 1954.

42. Betancourt, *Política y petróleo*, p. 669.

43. Vallenilla, *Making of a New Economic Order*, p. 71.

44. Creole Petroleum Corporation, *Annual Report, 1973* (New York), p. 11.

45. For the causes of the 1958 revolution, see Edwin Lieuwen, *Venezuela: A History*, 2d ed. (London, 1965), pp. 98–101; for documentary evidence on graft and peculation see Judith Ewell, *The Indictment of a Dictator: The Extradition and Trial of Marcos Pérez Jiménez* (College Station, Texas, 1981).

46. Ibid., pp. 103–105.

47. Interview, Caracas, July 7, 1962.

48. Ibid. This strategy was only partially successful.

49. Sampson, *Seven Sisters*, pp. 174–75, 188–89.

50. Tugwell, *Politics of Oil*, pp. 67–73.

51. Ibid., p. 53. One-third of the domestic petroleum market was awarded to the CVP, but it did not exercise this option until the late 1960s.

52. Ibid., pp. 54–59.

53. Interview, Albuquerque, N. M., Aug. 8, 1975; Tugwell, *Politics of Oil*, p. 35.

54. Creole Petroleum Corporation, *Annual Report, 1973*, p. 11. In this same period the Middle East increased its production from 4.5 million b/d to 7.5 million. See Vallenilla, *Making of a New Economic Order*, pp. 76–77.

55. Peter R. Odell, "The Oil Industry in Latin America," in *The Large International Firm in Developing Countries: The International Petroleum Industry*, by Edith T. Penrose (London, 1968; reprint, Westport, Conn., 1976), pp. 296–97.

56. Sampson, *Seven Sisters*, pp. 185–92.

57. OPEC, *Resolutions 1 and 2* (Baghdad, Sept. 14, 1960) (Caracas, 1961).

58. Adam Smith, *Paper Money* (New York, 1981), pp. 59–63.

59. Tugwell, *Politics of Oil*, pp. 86–96. Under the 1967 Hydrocarbons Reform Law the government's percentage share was again increased on service contracts. In addition, Shell and Standard were required to install desulfurization units on their refineries adequate to meet the new 1 percent sulfur content permitted in the United States market. See Vallenilla, *Making of a New Economic Order*, pp. 127–34.

60. Tugwell, *Politics of Oil*, pp. 103–111.

61. Sampson, *Seven Sisters*, pp. 250–57.

62. Ibid., pp. 268–72.

63. Donald L. Herman, *Christian Democracy in Venezuela* (Chapel Hill, 1980), p. 154; Tugwell, *Politics of Oil*, pp. 121, 129.

64. Tugwell, ibid., pp. 122–24. In June of 1973, Congress granted the CVP a monopoly on domestic market sales, effective by 1976.

65. Herman, *Christian Democracy*, pp. 152–53.

66. Franklin Tugwell, "Petroleum Policy and the Political Process," in *Venezuela: The Democratic Experience*, ed. John D. Martz and David J. Myers (New York, 1977), p. 246.

67. Betancourt, *Venezuela's Oil*, pp. 41–42. Although the abbreviation for Petroleos de Venezuela, S.A., was originally Petroven, some small businessman had the name registered and demanded a fortune for the rights; hence the change.

68. Venezuela, Oficina Central de Información, *Ley de nacionalización de petróleo* (Caracas, 1976).

69. Betancourt, *Venezuela's Oil*, p. 46.

70. Paul E. Sigmund, *Multinationals in Latin America: The Politics of Nationalization* (Madison, 1980), pp. 226–53.

71. The government-company profit split in the early 1960s was 62–38 in Venezuela compared with 55–45 in the Middle East; by 1970 it was 78–22 in Venezuela, 72–28 in the Middle East. See Tugwell, *Politics of Oil*, p. 150.

72. Ibid., pp. 5, 151.

73. Interview, Caracas, Jan. 4, 1976.

74. Ibid.

75. These works include: *The Monetary Question; The Trade Union Clause; Venezuela y su petróleo; Outline of a Policy; Oil: Essence of the Earth; The Tax Amendment; Public Expenditures and Oil; The Oil Pentagon; Oil Reserves; Service Contracts; Oil and Dependency; El desastre* (with Domingo Alberto Rangel).

76. Vallenilla, *Making of a New Economic Order*, pp. 111–19.

77. Juan Pablo Pérez Alfonzo y Domingo Alberto Rangel, *El desastre* (Valencia, Venezuela, 1976), pp. 213–45.

78. Venezuela, United Nations Mission, *Venezuela Now* (New York, 1976).

79. For fascinating details on these new operating companies and their different management styles inherited from the majors, see Gustavo Coronel (a former Gulf employee), *The Nationalization of the Venezuelan Oil Industry: From Technocratic Success to Political Failure* (Lexington, Mass., 1983), chap. 5 and pp. 109–113.

80. Ibid., pp. 92, 167.

81. Sigmund, *Multinationals in Latin America*, pp. 245–47.

82. Coronel, *Nationalization of Venezuelan Oil*, pp. 169–77.

83. Terry Lynn Karl, "The Political Economy of Petrodollars: Oil and Democracy in Venezuela" (Ph.D. diss., Dept. of Political Science, Stanford University, 1982), pp. 15–17.

84. Interview in Karl, "Political Economy of Petrodollars," 1976, p. 18.

85. Coronel, *Nationalization of Venezuelan Oil*, p. 166; Karl, "Political Economy of Petrodollars," pp. 221–22, 552–68.

86. Karl, "Political Economy of Petrodollars," pp. 24, 624.

87. Ibid., pp. 623–30.

88. Coronel, *Nationalization of Venezuelan Oil*, pp. 174–76, 219–32.

89. For the stormy history of the United States market for Venezuelan oil, see Stephen G. Rabe, *The Road to OPEC: United States Relations with Venezuela, 1919–1976* (Austin, 1982).

90. Coronel, *Nationalization of Venezuelan Oil*, pp. 215, 232; *The Economist*, Aug. 6, 1983, p. 61.

91. Coronel, *Nationalization of Venezuelan Oil*, pp. 181–94. As production from existing fields began to decline, PDVSA in 1979 decided to try to take up the slack from the FAJA, the huge bituminous tar belt north of the Orinoco River. Reserves there are estimated at one trillion barrels, 15 percent of which is recoverable with current technology. Development began in 1980, with production expected to begin in 1990 at 100,000 b/d. However, operations were suspended in 1982 due to the financial crisis. See Rafael Alfonzo Ravard, "La industria petrolera venezolana y su plano de inversiones, 1980–1990," in *El desarollo de Venezuela: Situación y perspectivas*, by Embassy of Venezuela, (Washington, D.C., 1980), pp. 41–47.

92. *Wall Street Journal*, Feb. 16, 1982, p. 31.

93. Coronel, *Nationalization of Venezuelan Oil*, pp. 226–27.

Alfred H. Saulniers

The State Companies:
A Public Policy Perspective

Of the top ten companies in the *Fortune* Foreign 500 list of firms for 1981, five were public enterprises.[1] On the basis of commonly accepted folk wisdom, one would expect that in comparison of the return to stockholders' equity, private enterprises always outperform public ones. In fact, almost the opposite occurred after ranking these ten companies by rates of return. The top three firms were all government owned and no government corporation earned less than 12 percent. These five government companies were all petroleum producers.

Performing the same calculations using 1982 data reveals the repercussions, on both private and public enterprises, of the slump in petroleum prices. For the largest oil producers, sales dropped 7 percent and net income plummeted by $10.4 billion, or 29 percent, compared with the previous year.[2] The best return on stockholders' equity among the top ten was by a nonpublic, nonpetroleum producer. Two of the petroleum producers lost money, and one of them, ENI of Italy, recorded a $1.2 billion deficit for a loss of over 70 percent of equity.[3]

In recent years and often in devoloping economies, confrontation on the issue of public ownership has arisen in part because the public petroleum producers have come to dominate industrial rankings. Their high and sustained profits have caused cash-strapped governments to depend increasingly on public enterprises for revenue and hard currency, while their investment needs have often occasioned government gloom. The profits of these corporations have forced a reevaluation of the conventional concept, "parasitic parastatals," while their great investment needs have coerced foreign and domestic suppliers into dependency relations. Their merely having high and sustained profits has exasperated antigovernment-interventionist critics, and their insatiable hunger for capital has encouraged those who favor dismantling the companies and converting them into private operations. Recently, however, profits have weakened in the face of the world oil glut, while their investment needs have remained strong.

This chapter examines several public enterprise issues in the light of the individual country cases presented above. In doing so I have put aside the usual emphasis of the public policy analyst about whether or not a public enterprise has attained its stated or implicit goals. Instead, the concern here is to analyze the public enterprise as a type of organization that is institutionally distinct from general government and that has particular institutional needs. Far from nurturing an ideological bias against public enterprises, I take as a point of departure that they exist (having evolved over many years), they have a role to perform, and they should be allowed to do it. To the extent that the state oil companies are in trouble today, my thesis is that persistent and incoherent government meddling is at fault, not any inherent defects in this form of enterprise.

To develop this line of argument, six major topics are addressed. These include: motives for public enterprise creation or continued existence; public enterprise goals and objectives; public enterprise transfers to the central government; central government transfers to public enterprises, styles of public enterprise management; and public enterprise relations to other government agencies.

The previous chapters by Esperanza Durán, Edwin Lieuwen, Carl E. Solberg, and John D. Wirth provide abundant and high-quality information with which to examine these six issues for the major Latin American public petroleum producers. Hence, each topic will be investigated by referring principally to the materials in those four case studies.[4] Two notable contributions are made in this manner. First, focusing on the formative years of the public enterprises offers useful insights into the problems of newly created government companies and, by extension, into the historical roots of many yet-unresolved problems. Second, since all oil-producing countries in Latin America have established and promoted their own public petroleum producers, an analysis of the influence of key individuals in company formation and development and of their approach to these six issues may elucidate the commonality of problem posing and problem solving.

Motives for Public Enterprise Creation

The literature on public enterprise origins is dominated by two main perspectives. One stresses what might be called the historical stages of development approach, while the other takes a taxonomic approach. Working from the individual case studies, one can readily demonstrate the shortcomings of both.

To explain how Latin American state company formation and growth proceeded through well-defined, sequential periods,[5] the historical stages approach postulates a logical series of consecutive motives that are supposed to have impelled governments to create different types of public enterprises during different time periods.[6] For example, until the first third of the century, Latin American primary-product, export-oriented economies are said to have maintained a public enterprise portfolio to meet the needs of the foreign-owned or -dominated exporters of agricultural or extractive products. Consequently, the state's major roles were to supply basic infrastructure, to form regional or sectoral promotion agencies, to mount rescue operations when private interests did not obtain a sufficiently high rate of return, and to make available public services such as adequate water supply and transportation networks.

A major motivational change is supposed to have occurred when the Great Depression and World War II cut off Latin America from its traditional European or North American suppliers of manufactured consumer goods and materials for low-level manufacturing processes. This opened the second stage, that of easy import substitution of foodstuffs, textiles, and any other industrial processes whose simple engineering technology and limited capital requirements were within the reach of domestic investors, with their penchant for conservative investments. While tariffs and quotas to protect the new domestic firms were the norm, occasional exceptions to the general pattern were found in politically based nationalizations of public services and basic transport, as well as rescue operations of foreign firms made unprofitable to their overseas owners by the expiration of government-granted concessions or by the imposition of government-imposed barriers to unfettered profit remittances. To be sure, George Philip notes that the main oil exporters of Venezuela, Colombia, and Peru suffered no prolonged effects of the depression and that only the oil importers, when faced with a balance-of-payments crisis, did not always have sufficient foreign reserves to pay for oil imports. For those countries in particular, the need to save foreign exchange became systematized later in the import substitution push.[7] Thus, the relationship between the depression and oil public enterprises is only indirect and tenuous. However suggestive, the approach stressing historical stages of development provides a broad framework that by itself offers only partial and at best very general insights into the formation of the oil companies.

The taxonomic approach to public enterprise formation systematically attempts to divide government objectives at the time of company creation into discrete public policy categories based on underlying political, ideological, or economic grounds. Since governments do in fact give a variety of reasons for creating public enterprises, classifying motives may be useful. Some governments preferred general public ownership for ideological reasons, while others have nationalized only those key industries previously held by foreign interests. Some companies entered the public portfolio following a revolution, and others were made public to capture the economic rents from natural resource exploitation. National security considerations dictated the public ownership of some firms, while yet others were rescued when their private owners went bankrupt. In practice, taxonomies may be simple (each category encompasses one basic motive) or compound (each initial subcategory is further subdivided), but in general there is little underlying rigor or consistent application of this approach. Actual categorizations differ widely.[8] In short, the taxonomic perspective also has its shortcomings.

Argentine President Roque Sáenz Peña created the Dirección General de Explotación del Petróleo de Comodoro Rivadavia in 1910 as a dependency of the Ministry of Agriculture principally for two preventive motives. The first, common in other sectors and across many countries, was a nationalist preemptive strike. He feared that American companies would come to dominate oil exploration if the government did not take a major direct role. Second, the president hoped to promote stability by insulating the economy and, in particular, government installations from the potentially disruptive effects of strikes by British coal miners. Thus, one explicitly stated objective was to produce and sell petroleum to the navy and the National Railways.[9]

During the early 1920s additional motives governed the operation and expansion of the now-renamed Yacimientos Petrolíferos Fiscales (YPF). First, national security concerns following the World War I–induced fuel shortages made army officers even more reluctant to depend on imported British coal. Self-sufficiency in petroleum became their watchword. Second, misgivings of influential civilian and military interests over foreign ownership of Argentina's oil led to their insistence on state ownership to guarantee adequate promotion of private manufacturing. Third, self-interest must not be discounted, as some military officers certainly envisaged an institutional role for the army, as well as personal upward mobility, in leading the new YPF. Taking a more limited point of view, Philip emphasized political over strategic, economic, or personal considerations. "The underlying motive . . . lay as much in Buenos Aires's distrust of the provinces as in the behavior, or even the presence, of the oil companies themselves."[10] However one weighs the evidence, it is abundantly clear that several motives were at play.

Although early Brazilian petroleum development bore many similarities to that of Argentina, it followed a somewhat more sinuous path. An underfunded and overbureaucratized program was established under the Ministry of Agriculture after World War I to carry out exploration activities. Its successor agency, the Departamento Nacional do Produção Mineral (DNPM), while embodying the military concern for national security, was given the same task. This in turn was succeeded by the Conselho Nacional do Petroleo (CNP). CNP was established in 1938 for a conjunction of internal and external motives. Domestic reasons included strategic fears of Brazil's vulnerability to any cutoff of imported petroleum supplies given the increasing likelihood of a war in Europe, developmentalist desires to boost industrial growth while controlling its direction, and nationalist hostility to the majors. External ones were the pragmatic response to the possibility of obtaining recently nationalized Bolivian crude through intergovernmental agreement and the ideological response to Enrique Mosconi and the Argentines.

Petrobrás was established for slightly different motives. The most important of these was to find and extract enough domestic petroleum to avoid a balance-of-payments constraint on industrial development, since by the early 1950s fuels and vehicle parts accounted for more than a fourth of Brazil's import bill. Almost equally stressed was the ever-present security consideration of overcoming the possibility of strategic interruption in the supply of imported crude. When created, Petrobrás was expected to be the financially independent segment of the state's petroleum operations, while CNP would set policy and rates. By the late 1950s, however, Petrobrás had taken over most policy-making functions.

Mexico too experimented with various institutional forms for its state petroleum corporation. In 1925 President Plutarco Elías Calles established the Control de la Administración del Petróleo Nacional (CAPN) to compete with the existing foreign-owned producers and refiners and to regulate domestic prices. Its successor, the 1933 joint venture Petróleos de México (Petromex), was to enlist the financial support of domestic entrepreneurs in meeting two additional objectives: maintaining a steady petroleum supply to the government and to the railroads and training local technicians. Due to a shortage of government and domestic private capital and the exclusion of foreign participation, Petromex failed to live up to expectations, and it was replaced by multiple, and at times overlapping, entities, often with undesignated or vaguely defined motives. In 1937 Petromex was transformed into the Administración General del Petróleo Nacional, which was complemented immediately after the expropriation by the short-lived Consejo Administrativo del Petróleo, which itself was succeeded in mid-1938 by two hastily organized companies: Petróleos Mexicanos (Pemex) and Distribuidora de Petróleos Mexicanos. This bewildering display of bureaucratic complexity resulted in duplication of functions, obvious overstaffing, and high labor costs, and a semblance of administrative unity did not come until 1940, when, to obtain better management while co-opting the labor union, all production, refining, and distribution were centralized under Pemex.

Motives for the 1938 expropriation and subsequent creation of Pemex were muddled. With the government's goals never made explicit and those of Lázaro Cárdenas never clarified, expropriation was widely perceived as nationalist self-assertion by dramatically expelling the exploitive foreign oil concerns. It was justified a posteriori as necessary to promote Mexican industrial development through judicious use of cheap oil; and it was viewed by labor as the creation of a bottomless font of patronage. Pemex, hastily organized to take over the fields, was shrouded in metaphors and thrust to the fore as the symbol of national sovereignty, the vindication of national development priorities, and the prime mover of industrial, technological, and manpower transformation.

Venezuela's Mining Minister Juan Pablo Pérez Alfonzo in 1960 fulfilled the long-postponed Acción Democrática attempt to promote economic independence by establishing the Corporación Venezolana de Petróleo (CVP).[11] Essentially a long-suppressed nationalist and populist reaction to the scandalous early multinational company–government interactions, CVP was an attempt to create a more activist state by providing it with a presence in the petroleum sector similar to that enjoyed by other Latin American countries.[12]

Petróleos de Venezuela (first called Petroven, but later PDVSA) was created in 1975 to explore and exploit the deposits of the nationalized private companies. It was established for preemptive purposes—to provide for early reversion to the state of the 40-year concessions expiring normally in 1983, thereby assuring that the fields would not be depleted, that plant and equipment would be maintained, and that production levels would stabilize. Following a long period of ever-increasing government control, regulation, and taxation, the anticlimactic nationalization benefited the companies by severing their adversary relation to the government over taxes, production, and technology and relieving them of the burden of dealing with militant labor. Unlike the bitter relationship that marked the Mexican takeovers, nationalization in Venezuela was relatively quiet, and the sales, management, and technology agreements meant that the interests of government and multinationals were intricately linked. The relatively quiet takeover notwithstanding, nationalization was both fostered and justified by the government in ideological terms using arguments based on dependency notions.[13]

For all the countries examined above, nationalization was not as sudden as commonly perceived. Rather, in all cases it was the culmination of a long-term movement. Precursors to the existing state companies were established years earlier; successive reorganizations and pressures to consolidate the government's position, often in the name of vague and all-encompassing national security or nationalistic considerations and only rarely based on improving efficiency conditions, resulted in the corporate forms by which they are known today.

What has been called the historical stages approach appears less useful in explaining why some of these companies were originally established than in justifying why they later grew. Consider, for example, the early state presence in the petroleum industry at the beginning of the century in both Argentina and Brazil during the export stage before the depression, when import substitution industrialization began in earnest. The taxonomic perspective may isolate nationalism as a major factor in company creation. But, while nationalism in the form of freeing domestic policy makers from the pressures of world market forces appears to have been behind the formation of many of the public petroleum producers, it was a nationalism often obscured by and entwined with a series of other motives, including price controls, previous experience with multinationals, the nature of the country's military formation, and reaction to manifold domestic pressures.

Certainly, it may be said that ignorance, uncertainty, governmental self-interest, and fear of the future marked the beginning of the petroleum role of the state in Latin America. If ongoing company reorganizations are any indication, ignorance and uncertainty may well be intrinsic to Latin American public enterprises in general. Justification for many early actions today appears murky, shrouded in obscurity, and, with the exception of Venezuela, more a hedge against potential transnational behavior than a reaction to either the past or the recent economic or political activity. For Argentina and Brazil, the importance of getting a toehold in a new industry should not be discounted, while for other countries the demonstration effects of the earliest public petroleum enterprises were important. Creating and maintaining a successful state oil industry required abundant technical sophistication—Wirth touches on this in his introduction. It happened that this was easier to fulfill than another fundamental requirement: clear and consistent political direction, a factor that has always been critical to the relative success or failure of the Latin American public enterprise.

Goals of Public Enterprises

Measuring the performance of public enterprises is a common undertaking in developing countries. In carrying out what apparently should be a simple exercise, however, the issue of how to assess public enterprise performance is often neglected. Given that public enterprises are a hybrid combination of a legal corporate structure overlaid with certain government attributes and also given that public enterprises are often expected to pursue social as well as commercial goals, it is clear that broader issues than the customary analysis of income statements and balance sheets need to be addressed. If public enterprises must attain social goals and respond to the needs of various constituencies, then they should be judged, at least in part, according to these objectives. But unless all goals are made explicit, there is no clear basis against which results may be judged. Further, without precise goal specification commercial losses can all too easily be imputed to meeting social objectives, thereby masking the effects of poor management. This section will not evaluate goal attainment; rather, it will focus, within the framework of multilevel public enterprise duality, on how the mix between social and commercial objectives is determined.

In investigating the origins of public enterprise goals, particular attention will be paid to the dichotomy between the external goals, which are fixed outside the firm by the government and which attempt to constrain management to act predominantly in the national interest, and the internal goals, which comprise the yardsticks that company management defines as criteria for success.[14] The often-addressed accountability problem that deals with the design of adequate control and monitoring systems to ensure compliance with external goals has neglected two important aspects of these petroleum companies: externalization of internal goals when company executives appropriate national executive prerogatives; and heavy-handed government involvement in company affairs, when internal goals are subordinated or nonexistent.

External constraints on Argentina's YPF varied over time. In contrast to President Hipólito Yrigoyen's lukewarm support during his earlier term, President Marcelo T. de Alvear made YPF's development a top priority of his administration. To this end, he generally supported Enrique Mosconi, but with little economic backing. While according to a 1923 law, the president could name YPF's director-general and its board of directors, the company was given total operating autonomy except for annual ministerial budget approval. In both cases internal goals predominated, under Yrigoyen from neglect and under Alvear from deliberate policy.

Under the presidency of Gen. Agustín Justo, internal and external political factors strongly influenced petroleum policy. While the policies arrived at were clearly compromises, their net result was consistent: YPF's autonomy was circumscribed, the movement to a monopoly was derailed, and its profits were taxed away. These measures considerably so weakened YPF both administratively and financially that even with the increased territorial reserves after 1935 the company's exploration and production were held to suboptimal levels.

The battle for the lucrative Buenos Aires market also exhibited this pattern of increased priority for externally imposed goals at the expense of management-defined objectives. Uniform pricing provided a strong incentive for the large foreign firms to concentrate on the Buenos Aires market, but confining YPF to operations in the rest of the country would not have been profitable. Under Mosconi, agreement with the municipal authorities favored YPF in the allocation of retail outlets. Under Justo the appointed mayor, a former Standard Oil lawyer, reversed this policy. By the late 1930s, a compromise guaranteed YPF a stable, but minority, market share. This agreement lasted a decade. George Philip summed up government-YPF relations quite succinctly: "For every Argentine government since 1930, the needs of YPF were considered to be at best secondary to other political or economic objectives."[15]

Brazil's CNP was mandated to develop Brazilian self-sufficiency by exploring, marketing and refining petroleum products. In doing so, it was to insulate the country from further balance-of-payments problems and to minimize the potentially disruptive effects on the economy of impending perturbations in the supply system. Another highly controversial objective was to gradually preempt and circumscribe foreign firms, with the first steps being the regulation of the national market and the control of national energy resources. Although CNP achieved a high degree of operational autonomy under Júlio Caetano Horta Barbosa, it formulated no long-term, unrestricted investment policy.[16] CNP was subject to yearly budgetary outlays and treasury review, ostensibly for the purpose of monitoring goal compliance but with the effect of undermining planned company efforts because of consistently tight budgets. Consequently, little money went into exploration, and most of that into vindicating CNP's credibility by improving the small Bahia field that was discovered in 1939.

Later, Petrobrás was charged with moving the country toward self-sufficiency by profitably refining oil and channeling those profits into a major exploration campaign. It was permitted a wide degree of autonomy and largely operated outside actual budgetary control. Although management realigned these priorities in the early 1970s to stress profits over self-sufficiency, it was overruled by President Ernesto Geisel and top ministry officials who stressed finding domestic oil even at high cost. As the situation worsened, risk contracts were signed with foreign firms to try to end the balance-of-payments and oil deficits.

Mexico's Pemex has always been handicapped by strong and conflicting external signals about the nature of company goals. From its inception, presidential control was exercised through naming the members of the board of directors and through the requirement for annual budget approval. While the government board members tried to steer Pemex into a businesslike orientation that would make it financially independent of the government, the labor representatives insisted that the company be run principally to protect the workers and to promote their well-being. Both groups, however, agreed on two key and quasi-autarkic objectives: Pemex should provide at least enough petroleum to meet domestic demand, and it should support with subsidized energy prices the goal of independent development via import substitution. Later, the 1941 Petroleum Law, which asserted executive political control over company decisions relating to exploration, production, transportation, refining, storage, and distribution, left a debilitated management to wrangle with a relatively strong union.

During the 1960s, Venezuela's CVP was assigned long-term objectives of participating in all facets of the oil industry, including exploration, production, and marketing. In the short term its main concern was to increase the income reverting to the state by instituting competitive bidding on long-term service contracts for national reserve parcels.[17] CVP's market share goals were set by central government authorities. In 1964, Decree 187 set a target of 33 percent of the domestic market by 1970, and Congress subsequently passed a law giving it 100 percent of the domestic market by 1971; however, private competitors' passive resistance to CVP's obtaining distribution outlets led to missing both targets.[18]

Relations between PDVSA and the Carlos Andrés Pérez administration remained cooperative and cordial from 1976 to 1979. PDVSA (Petróleos de Venezuela, S.A.) was given two main tasks at first: to provide for a smooth transition from private to public enterprise and to redress the deliberate neglect of previous private management. The latter goal included modernizing obsolete plants and equipment, undertaking long-postponed explorations, and upgrading the caliber of local management. To ensure meeting these goals, President Pérez named the members of PDVSA's board on the basis of technical criteria, including management expertise and oil industry experience. The minister of energy initially cooperated by maintaining a hands-off policy. By 1979, the petroleum bonanza proved too tempting a morsel for the central government authorities to resist; board members' terms were halved, Energy Ministry approval of PDVSA's budgets was required, and the company was increasingly criticized for its high costs and generous employee compensation. In 1981 political criteria had replaced technical ones in determining board members, and direct ministerial interference became common.

Argentina's Mosconi effectively set YPF's clear and ambitious goals during the 1920s. These included purely internal concerns (reorganizing and expanding YPF to compete with foreign firms), externally motivated internal change (price reduction through vertical integration to stimulate industrial growth and to court the average consumer), and external change for both internal and external reasons (promoting a joint-venture monopoly to boost efficiency and to satisfy nationalism). Even provincial opposition, political antagonism, and constitutional questions did not deter Mosconi from trying to force YPF to a position of preeminence.

Horta Barbosa decided on a strong production role for CNP and, after bringing in the Bahia field, followed the pan-Latin trend begun in Argentina and Uruguay. He emphasized refining to control local supplies and pricing and to finance exploration. President Getúlio Vargas generally supported Horta Barbosa's plans, but not to the extent of granting CNP the monopoly later granted to Petrobrás. Horta's plan to expand CNP into a national energy council was bitterly opposed by coal interests and was buried.

While oil was cheap, Petrobrás downgraded its primary mission of domestic self-sufficiency in the interests of commercial considerations and became instead an extremely profitable full-service company with important overseas operations. The high priority accorded to internally set objectives could not last forever, and management was overruled by President Geisel by 1978 in response to external market and balance-of-payments pressures. The by-then-institutionalized hope of controlling the energy policy of Brazil was thwarted when Petrobrás failed to get approval to control coal and alcohol production in the 1980s.

In Mexico, control over internal goal setting was denied Pemex almost from its beginning. According to the 1941 Petroleum Law, all major decisions were to be made outside the company by the executive branch of government for political or social, not for economic or business, reasons.

Although some central control is always needed for a public enterprise system, too often the four governments in question, in the name of macroeconomic planning or political considerations, have usurped a decision-making power over matters that properly are the prerogative of management. To make matters worse, targets for their public petroleum producers were rarely clearly defined or adequately stated. Instead, imposed goals were often vague, mutually contradictory, contrary to management wishes, or overwhelmingly politically motivated. Only rarely, as with Mosconi's early years, was a strong company leader supported by the nation's chief executive able to take charge and set ambitious technical goals that were later externally ratified.

Except for Petrobrás, in the few cases where management autonomy was allowed to flourish, such freedom was short-lived. Consequently, the wide policy swings from autonomy to heavy-handed control, which plagued YPF's first ten years and PDVSA's first five, impeded the development of coherent public enterprise direction and often destroyed management morale.

Transfers from Public Enterprise to Central Government

Common wisdom has it that transfers from public enterprises to central government are negative and that transfers flow only in the opposite direction. Since public enterprises are often found in capital-intensive sectors of the economy, their capacity to absorb investment and other funds is quite high.[19] Further, and again according to common wisdom, since public enterprises invariably end the year in deficit, they simply have no money to transfer to the central government. Instead, the only remaining issue is how much additional, extrabudgetary funding must be found by the government to bail out the companies in its portfolio.

The case studies presented in this volume cast considerable doubt on this folk wisdom. During the 1970s, and in part as a result of OPEC-inspired price rises, the situation changed dramatically in favor of the petroleum companies. It is apparent from even a cursory study of public enterprise history, however, that central government authorities had decided long before OPEC to use their petroleum enterprises to raise badly needed revenue. This section will focus on three key decisions that must be made in deciding the disposition of public enterprise surplus: profits, taxes, and prices.

Public enterprise profitability, per se, will not be examined in this section.[20] Instead, control of the profit-disposition decision and the magnitude of profit transfers will be stressed. The two are interrelated and bear on many other aspects of management behavior. Other things being equal, the greater the central government take, the greater the incentive for company managers to inflate operating costs, boost employment, approve questionable investments, and pamper themselves with personal perquisites—all of which result in lower profits. A high government take coupled with extensive government interference in internal company affairs will usually be associated with extremely bureaucratic management styles, a high rate of top executive turnover, excessive delay of needed investments, and lower profits.

Argentine President Yrigoyen permitted YPF to control its own profits when budgetary support of oil development ended. Mosconi then predicated his expansion plans on the assumption that increased production-induced profits would finance investment. Although unanticipated production drops during the late 1920s led to falling profits-to-sales ratios, profits in 1929 totalled more than all central government transfers from 1910 to 1916. Under President Justo, the major profit-shifting mechanism changed from outright transfers to taxes. Profits disposition was also a contributing factor to the YPF provincial rights dispute, since a major fear of the Salta oligarchy was that profits would be funnelled out of Salta to support a bloated porteño bureaucracy.

After 1953, Petrobrás was able to generate profits by refining imported crude. The company controlled the profit-sharing decision and plowed this refining income back into operations. After the

1974 price rise, however, excess profits had to be transferred to the treasury.

Mexico's Pemex never controlled the profit disposition decision. Because its profits went directly to the Secretaría de Hacienda, the company was continually hampered for lack of funds. Further, key profit determinants were beyond management's control, as both final prices and wages were politically set by central government authorities; this contributed to a lack of management concern about efficiency. In spite of government legal requirements, over the years much of the operating profit has been distributed to the workers in the form of annual bonuses.[21]

In Venezuela during the late 1970s, PDVSA's profit increased as a result of the petrodollar boom and accounted for three-fourths of total government revenues. To guarantee that PDVSA would have sufficient capital to accomplish its mandated tasks, the 1976 nationalization law established a development fund to which 10 percent of yearly export sales revenue would be contributed for exclusive PDVSA use. By the end of 1982 President Luis Herrera Campíns dispossessed the company of its investment funds to rescue the government's ailing finances, cancelled investment projects, and earmarked $2 billion of company money to subsidize the bankrupt and corrupt labor-owned Banco de los Trabajadores.

The most common method used by central government to capture the available public enterprise surplus is an income tax. Rates tend to be high, which may lessen management concern for cost minimization.[22] Taxes on the volume of production, usually in the form of royalties, are also levied. Initially assessed in the form of fixed amounts based on area or surface exploited or on the volume of oil extracted, such royalties have in recent years been calculated on the value of production. Additionally, sales or excise taxes have been charged on products sold domestically. The case studies demonstrate an ample tax mix.

In the 1930s under President Justo, YPF was weakened by taxes to meet Argentine debt payments. Since then, YPF has consistently been regarded as a convenient revenue source to ease Argentina's chronic and serious deficits. The cash-starved depression years of the 1930s witnessed an ever-increasing tax burden. Besides the income tax, provincial and federal government royalties were levied, and the sales taxes charged by the provincial governments were absorbed by the company to maintain a uniform national price. By

1935, the present tax structure was in place, but the rates continued to climb. Income taxes originally set at 10 percent in 1932 rose quickly to 30 percent the following year and by 1975 had reached approximately 60 percent. By then taxes were earmarked for highways, railroads, and electricity.

During the 1930s, YPF transferred, in taxes and lost revenue alone, an amount equal to its total earnings during the previous decade, and the average government take during the 1930s was 50 percent of gross earnings. Since YPF's investment in exploration and field development was held to suboptimal levels, the net effect of this transfer was that Argentina never achieved petroleum self-sufficiency.

Although the newly formed company was expected to pay taxes, Pemex increased its expenditures so dramatically following the expropriation that it suffered a decline in net earnings of 36 million pesos, leading to substantial losses. It ceased payment of taxes and was bailed out the next year by a treasury transfer to pay tax arrears. The 1941 Petroleum Law that reorganized Pemex as a public utility under the exclusive jurisdiction of Mexican federal authorities gave them the right to tax the company. The tax burden, when measured in current pesos, grew steadily, but at a decreasing rate, until the late 1960s. The 1970s witnessed tremendous changes. By the end of the decade, Pemex paid sales taxes of 16 percent and a tax of 50 percent on crude oil exports.[23] Aggregate taxes paid by Pemex swelled from 3 percent of Mexican government tax revenues in 1971 to more than 12 percent in 1979, which was more than the total from *all* private industry combined.[24]

The Venezuelan government continued the policies set before nationalization and relied heavily on taxes as its major means of capturing the economic rent from the petroleum sector. During the first two years of PDVSA's operation, the central government share of gross income, based on income tax, royalties, and profit-sharing mechanisms, exceeded 75 percent.[25]

Pricing policy is related to the first two mechanisms of surplus transfer, since it directly influences the amount of profit and net taxable income. As a way of exacting the surplus, however, it is the most politically sensitive because of its direct consumer effects. Ambivalent pricing policies have often led to public petroleum enterprises being squeezed, on the one hand by ever-increasing government demands for transfers and on the other hand by low fixed prices set outside the company. In some cases, the pricing decision was given such priority that public enterprises showed year-end, government-induced deficits.

Prices in exporting countries have often been kept low as a means of spreading the wealth, favoring the poor, and promoting industrialization. Such arguments, while conceived from a populist viewpoint, often have the opposite effect of leading to greater income concentration because of the strong correlation between consumption of petroleum products and income.[26] Below-market prices often result in an open-ended subsidy policy that acts as a permanent treasury drain and, given the regressive nature of taxation in many countries, contributes still further to worsening the income distribution. Another negative consequence of long-standing price distortions stems from their having contributed to building an imbalanced industrial structure based on erroneous market signals to investors and managers. Because petroleum product prices did not correctly reflect their scarcity, energy-intensive, rather than energy-conserving, industries had been nurtured until the withering effects of the 1970s petroleum crises.

Argentina's Mosconi set YPF's prices according to predominantly political motives, although they had strong economic and financial impact. By lowering fuel prices in 1929 he launched two YPF policies: establishing uniform national prices and maintaining them below world market levels. The oligopolistic nature of the Argentine market meant that YPF assumed a policing function, and all companies were forced to match its prices, which in turn resulted in a profits squeeze for all firms in 1930. While low prices undoubtedly aided the industrialization effort, as well as boosting the transport and construction sectors, the negative effects of this pricing policy on the company and on Argentine industry still can be seen today in the YPF income statements and the country's import balance.

Brazil's price-setting policy was influenced by the Argentine example. Horta was able to convince state governments not to build refineries with the explicit goal of channeling profits into the regional economies. Instead, he pushed through a plan to establish uniform national pricing to finance regional development and to underwrite the federal highway system. This long-lasting policy assured a high degree of domestic price stability.

During the late 1960s and early 1970s, Petrobrás astutely aban-

doned its price-setting ability and adopted the behavior of a dominant firm in an oligopoly. By allowing overall prices to rise high
enough to provide a reasonable return to the small and inefficient
refineries, it was able to earn large profits from its more efficient
operations.[27]

Pemex never had a pricing policy. Before nationalization the gasoline prices were set by the Mexican government as a basic commodity, while those of other petroleum products and oil derivatives
were fixed by the foreign companies. As stated above, the basic objective in postexpropriation government price setting was to keep
fuel costs as low as possible to promote industrialization. In fact,
that pricing policy subsidized both personal and industrial consumption at the expense of Pemex' financial health. During the late
1960s and early 1970s, the necessary financial resources for Pemex
to undertake significant exploration and development were not
available, since the domestic petroleum prices had been frozen for
fifteen years preceding 1973 and recourse to foreign financing was
unthinkable. It is not surprising, therefore, that after 1966 Pemex
had to import substantial quantities of crude and refined products.
With the lifting of the price freeze, the Pemex exploration and development budget increased 60 percent.[28]

Sharing the net revenue generated by petroleum exploitation has
often led to conflict between central governments and their public
petroleum enterprises. The governments, in their dual role as
owners of the petroleum reserves and owners of the company, often
expect to capture most of the net revenue by restricting the
companies' retained earnings to levels sufficient only to maintain
an ongoing flow of funds. The companies typically seek to amass
sufficient reserves to explore and develop new fields and to assure a
considerable return to managers for their highly specialized technical and managerial skills.

On balance, the governments have exercised sufficient power to
dominate the profit-sharing decision, even, as in the case of Venezuela, resorting to the confiscation of funds. The approach to taxation
has been relatively unsophisticated. Governments have been concerned only with using the companies as siphons for funds, rather
than with designing a taxation system that would be mutually beneficial. Production- and income-based taxes have been used indis-
criminately, without regard to the consequences. Production-based
taxes generally mean earlier and steadier revenue flows, while in-

come-based taxes lead to widely fluctuating income that is higher on average. The latter also stimulate higher production from marginal fields.[29] The price-setting decisions unfortunately have relied almost exclusively on short-term political input or on long-range economic considerations that were based on erroneous assumptions.

Transfers from Central Government to Public Enterprises

While folk wisdom holds that unilateral transfers flow from governments to public enterprises, experience with state petroleum producers almost invariably has demonstrated the opposite. The rare transfers to the companies usually resulted from misguided government policies. Although the price of petroleum products was often held down under the populist assumption that they were basic consumption goods, that assumption was erroneous. In fact, as consumer budget surveys have shown, petroleum derivatives are mostly used, even including all secondary effects, by a small upper-income section of the typical Latin American country's population. Trickle effects notwithstanding, government subsidies through the public petroleum producers to the highest income earners do not redistribute income more equitably. The sounder policy of promoting a viable industrial mix by pricing gasoline as a true luxury item, while holding the price on other petroleum derivatives close to world market levels, has rarely been followed, except in Brazil.

Alternatively, transfers for investment purposes or to provide initial capitalization for a company, when the pricing mechanism was properly functioning, could have served the useful purpose of meeting a rising demand domestically rather than relying on imports. Instead, governments usually starved their petroleum companies for cash or grudgingly transferred the promise of later capital infusion through minimum cost mechanisms.

Argentina's capital-starved precursor to YPF attempted oil field development using funds obtained through budget allocations. Before 1916, these averaged only $500,000 per year and, while they were barely sufficient to produce a trickle of oil and to undertake minimum exploration activities, they could only be obtained over continued private sector opposition. Yrigoyen ended these transfers, and under fiscally conservative President Alvear, the capital shortage continued through the 1920s, in spite of YPF's steadily dropping production level and market share. This forced Mosconi to consider alternative financing sources, especially the creation of a joint public-private firm, to realize his ambition of making YPF a truly nationwide monopoly, a dream that never saw fruition.

During the 1940s, CNP activities suffered from grave financial difficulties, as tight budgetary control at the federal level meant that little money was available for financing company projects or increasing company equity to the point that it could carry out projects independently. CNP was also low on the priority lists for obtaining foreign funding. Only in 1949 was there a major capital infusion to buy a tanker fleet and to begin construction on a 45,000-barrel-per-day refinery, but even continuing budget allocations to complete these projects were subject to the uncertainties of the yearly budget process.

When Petrobrás was created, it was planned as a well-capitalized, financially independent company. Direct government capital infusion was minimal. CNP assets, including the tanker fleet, two refineries (one still under construction), and drilling equipment, were transferred to the new company. Most of the planned capital increase was to be raised gradually from the product of earmarked taxes such as a higher gasoline tax, taxes on luxuries, taxes on the private refineries, and import duties on refined products coming for the private distribution networks. There was also a mandatory stock subscription by the owners of private vehicles.[30]

The Mexican government only occasionally has had to transfer money to Pemex. In 1940, the government covered the company's tax arrears. On the heels of Pemex' near bankruptcy in 1958, following the Dwight D. Eisenhower administration's curtailment of oil imports to the United States, its debt to the state was restructured by converting it to 99-year 8 percent bonds. Notwithstanding this effort and an increase in petroleum-product prices, the situation continued to worsen, so in 1960 Pemex requested a debt moratorium on payments to the government.

Venezuela has consistently seen sizeable transfers flow from its petroleum producers to the central government. The change from private to public ownership led to no change in resource flows.

Public petroleum producers have regularly subsidized the activities of consumers, the inputs of most sectors of the economy including industry, transportation, and construction, and the coffers of central, regional, and provincial governments. Only rarely have company activities been subsidized either to cover extraordinary investment needs or to provide badly needed cash flow in the face of lost markets when local price adjustments were denied. Instead, non-market-based pricing mechanisms coupled with transfer of most if not all of the remaining surplus to government authorities through various mechanisms has resulted in the public petroleum enterprises being used as milk cows for the rest of the economy. Again, Petrobrás has been the exception.

Nature of Public Enterprise Management

Competent management, which is so essential for efficient public enterprise performance, has two preconditions: appropriate management skills and a performance-based reward structure. Low pay, limited autonomy, shortage of local talent, and promotion based on nepotism or political patronage increase the probability of unskilled management. Ample responsibility, abundant recognition, and adequate remuneration attract and motivate managers, while executive autonomy and performance bonuses reward quality administration. However, a comprehensive management system that places a premium on institutionalizing efficiency never emerges overnight. Not only during the crucial early years of the company, but also at other times when the preconditions for competent management are not met, rational and scientific decision making may not be found. Indeed, whenever a technically and administratively successful modus vivendi does not exist between government and its petroleum producer, the latter is especially prone to administrative chaos superposed on a technically talented substructure.[31]

So scandalous was the management of the Argentine state oil works during and immediately following World War I that it led directly to YPF's creation. Mismanagement, financial shortages, and labor disputes were the norm rather than the exception. Solberg concluded that the inefficiency was so great that it was "seriously discrediting the whole concept of a government-owned oil company."[32]

Under Mosconi, management was tightened and the worst excesses were brought under control. Mosconi's public sector entrepreneurial style was a blend of organizational genius, skillful public relations, arrogance, impatience, obstinacy, and self-righteousness. He was relatively open in seeking expertise wherever it could be found, and in fact much of the early technical YPF work was carried out by non-Argentines, but Mosconi's "scientific" decisions were taken by fueling rivalries among competing intracompany groups and favoring that group whose recommendations most closely fit his preconceptions. Although Solberg provides evidence that Mosconi viewed YPF (apart from himself of course) as not very efficient and excessively bureaucratic, a view that encouraged his scheme to inject private management skills by creating a mixed monopoly, others have attributed the notable YPF successes of the twenties to Mosconi-instilled organizational discipline.[33]

In 1932, the Justo government stripped away much of YPF's earlier autonomy and placed it under close government supervision. All major investments and all key operating decisions needed cabinet approval. These included new equipment purchases, new construction, loan requests, price setting, and provincial contracts. Thus, bureaucratic decision making became the government-enforced norm.

Later, management became extremely politicized. In the purges before Juan Perón's election in 1945, almost 70 percent of senior executives were removed for political reasons and, according to Philip, party "loyalty appeared to replace knowledge of the industry as a criterion for appointment."[34] The tendency to management inefficiency was exacerbated by the destruction of YPF's technical library, the suppression of technical information, and the cancellation of training abroad. Furthermore, YPF was unable to refurbish its worn-out equipment because of United States–imposed restrictions on exports to Argentina of drilling equipment, refinery apparatus, and repair parts for the petroleum industry.[35] Inadequate human capital, the extreme deterioration of physical capital, and the proliferation of bureaucratic bottlenecks led to near stagnation until the end of the Perón regime.

From the mid-1950s to the present, YPF has been marked by worsening trends in management. The ever-increasing government share of the sale price has led to severe capital shortages. Inadequate accounting procedures have reflected the lack of concern with cost considerations. Manifestly unsuitable people have risen to top positions on the basis of seniority alone. YPF's history has witnessed a disintegration into a deeply rooted and all-pervasive bureaucratic management style that is largely the result of decades of government-induced weakness.

In Brazil the DNPM was often criticized, particularly by a biased Monteiro Lobato, for being an unresponsive bureaucracy. Although the same charges were later levied against the CNP, one of its major accomplishments was to bring in the Lobato field, where oil was said not to exist, which may indicate a measure of technical competence. When seeking to increase management independence in determining key personnel and industrial policies, Horta hammered home the point that private enterprises as well as public ones were subject to bureaucratization.

During its early years, Petrobrás' conservative administrators were expected to maintain a low political profile for the company. They often, as did Mosconi in YPF, hired foreign experts as technicians and for training local staff. Later, this style changed to a more politically conscious one, where management generated a political constituency for the company and where, after a time, management factions and workers counted on external political power bases to support their positions in intracompany squabbles.[36] Following the post-1964 shift in relative power to the technocrats, a strictly administrative, nonentrepreneurial management style, similar to that found in large multinationals, emerged. Thus, Petrobrás today continues to be subject to the never-changing criticism of having a swollen, closed, semiremote bureaucracy that still prefers company expansion over all other objectives.

In Mexico, during Pemex' early years management never adequately developed. When most foreign technicians and senior administrators left immediately following the expropriation, the Sindicato de Trabajadores Petroleros de la República Mexicana assumed administration of the companies. While possessing the technical expertise to run the industry efficiently, its control was used simply to promote the economic well-being of the union leadership and to build patronage structures for them. The ensuing scandals over featherbedding and high-cost operations, while leading to reorganization under a mercantile legal form, did not endow company management with decision-making authority. Rather, the 1941 Petroleum Law removed all real autonomy from the company, authority that was only partly recovered much later under the astute directorship of Antonio J. Bermúdez. Even after this relative power shift Pemex still remained under firm political control.

Venezuela's shift from private to public enterprise management was not accompanied at first by a shift in management structure or styles. The previous corporate organizational structures were maintained under a PDVSA holding arrangement. The climate of confidence created by having production, sales, and consulting contracts signed with the former parent companies meant that their managers and technicians overwhelmingly remained with the newly nationalized industry. Later, management became increasingly frustrated by government meddling in company affairs, but any resultant shift in management styles is too early to detect.

The cases presented above illustrate that public petroleum enterprises are often technically qualified and ably staffed, but enmeshed in administrative chaos. Qualified technicians are not necessarily qualified managers, however, and with few exceptions the companies have not been able to negotiate for the autonomy necessary to exercise efficient management skills. During the early years a strong charismatic management style often emerged, blending the technical qualifications of an oil specialist with the compromising ability of a skillful politician. Venezuela's Pérez Alfonso is the prime example. Later generations of managers have evolved inbred, institutionally marked styles ranging from YPF's bureaucratic ossification to Petrobrás' multinational administration.

Three preconditions to shifting these companies to an efficiency-promoting management style are: central government recognition that problems exist, government consensus to implant adequate management structures, and government delegation of autonomous management responsibility to the companies themselves. Without these key changes outside the companies, quality management within them will not occur.

Institutional Relationships to other Government Entities

Other supervisory matters, besides taxation, pricing, and profits, include personnel, budgets, and strategic planning. At stake is the balance between central government's need to coordinate and control its public enterprises and company needs for autonomy in action. Effective control must not only avoid stifling management's innovative capacity in the name of regulation; it must also protect management from the possible disruptions caused by an unsympathetic or uncomprehending bureaucracy. Autonomy must be carefully balanced against accountability and tempered with a strong dose of protection from the government.

Public enterprises are commonly enmeshed in two systems, one involving central government institutions and the second with other public enterprises. Companies are subordinated not only to their controlling ministry or institution but also to other different central government agencies that exercise a controlling function. Incompatibility among the controllers' functions, policy disagreements among supervisory agencies, struggles over bureaucratic territory, frictional interaction with other publicly owned firms, or any other sources of contention may have strong repercussions on public enterprise profits or social yardsticks. Public petroleum producers, by virtue of their size, the political importance of their output, and the short-run inelasticity of consumer demand for their products, have been especially prone to opposition.

This dual systemic approach is generally not considered when elaborating coordination and control measures. Too often the accountability issue is limited to specifying top-down relations from the central government authorities to a specific company. This neglects the ever-present public enterprise interaction with other government-owned corporations. For example, two such companies may be found in a bilateral monopoly relationship, where one is the sole producer and the other the sole purchaser. Economic theory does not provide easy answers for the pricing and quantity decisions that have to be taken in such a situation, nor do governments. Companies failing to honor contractual arrangements may transmit their financial problems to the entire public enterprise system if allowed to go unchecked. Faulty relationships between the public petroleum producers and their governments and between the public petroleum producers and other public enterprises have been commonplace.

Mistrust by the Argentine Congress was responsible for the severe undercapitalization of the state oil commission under Yrigoyen.[37] Later, YPF's severe cash flow problems of the early 1930s were directly traceable to intercompany problems, since after 1929 other public enterprises, particularly the railroads, did not pay for the petroleum products received. Faced with the possibility of YPF bankruptcy, in 1932 bonds were issued to cover less than half of the estimated debt these firms owed YPF. These half-measures were exacerbated by forcing YPF to continue an earlier policy of discounting all its sales to other government agencies.

This problem continues, as major cash flow difficulties today arise from the failure of other Argentine public enterprises to pay YPF for fuel. Arrears on accounts receivable of up to ten years are not uncommon for sales to other government units, which account for 20 percent of YPF's sales. Further, settlement of the debts has been accomplished in the past through cancellation certificates, which are only valid in settling intragovernmental debt and useless in meeting YPF's cash needs.[38]

Brazil's CNP not only generally lacked support from major state agencies, but many of its dealings with them were marked by bureaucratic infighting over territorial jurisdiction. When founded, CNP had no control over hiring and firing, which was under the jurisdiction of the Civil Service Agency, DASP. It was not until several months later that Horta Barbosa exacted a promise from President Vargas to support autonomy in hiring. For more than a year after its creation, CNP struggled with DNPM over oil drilling, before wresting control away. In mid-1941, CNP's remaining development funds under the 1939–43 development plan were reallocated to buy aircraft. Shortly thereafter, in 1942 as a result of war-inspired power shifts, Horta lost control over all major policy making to the Coordinator for Economic Nationalization. No longer able to determine exploration, refining, fuels allocation, or pricing, CNP was downgraded to implementing policy set elsewhere and enforcing the unpopular wartime gas rationing.

Although Petrobrás was able, by virtue of its exceptional operating autonomy, to minimize the negative effects of intragovernmental relations, it had the standard commercial problems with other government agencies. When it began distribution of refined products in 1961, those agencies, which had hitherto been in arrears to the private distributors, simply transferred their poor business practices to Petrobrás.[39]

Likewise, in Mexico, commercial relations between Pemex and the other public enterprises were not always cordial. In 1947, nonpayment by other nationalized industries, especially the railroads, led to cash flow problems so severe that Bermúdez threatened to cut off supplies unless accounts were settled.[40]

The public accountability approach to public enterprise management seeks to devise adequate corrective measures to overcome deviations from government-imposed policy. The company level approach seeks to discover ways to promote internal efficiency. Both are based on the assumption that corrective actions, once implemented, will achieve their intended targets.

Both approaches fail to consider that being enmeshed in two relational systems, one with various government agencies and the other with many public enterprises, make many efficiency-related issues exogenous to company managers or their supervisors. Neither top-down directives to improve efficiency nor the best efforts of management to streamline company operations can solve the problem of other government customers failing to pay their bills. Thus, neither approach alone, nor both together, can provide the kind of policy prescriptions necessary to solve system-level problems. Only a broader strategic systems approach, which incorporates methods of dealing with the multiple relational systems sketched out above, can provide the basis for effective solutions.

Conclusions

This analysis of the origins of the major Latin American public petroleum producers has established that many of their current adversities are rooted in their early and checkered histories. Problems that arose during a company's formative years were frequently magnified with the passage of time or worsened through deliberate government policies toward public enterprises. Prominent figures in company formation or the technocrats who for a period ruled Petrobrás were often able to keep government authorities at a respectful distance, thereby upholding institutional integrity. After their tenure, however, public petroleum enterprise objectives were forcibly replaced with government ends.

Governments generally have been unconcerned with the problems of public enterprises as institutions and instead have viewed them as means of executing a particular, often transient, sectoral policy or as sources of badly needed revenue. The failure of governments to establish a national consensus on public enterprises deprived the firms of a relatively fixed and stable environment and subjected their operating milieus to the vagaries of political, regime, ministerial, or bureaucratic changes. Governments rarely tried to solve systemic problems that arose because public enterprises are involved in a closely interrelated structure; rather, they considered them in isolation, each one created for its own specific purposes—and beset with unsolvable problems. Governments generally did not attempt to streamline public enterprise supervisory systems, and instead have substituted overlapping oversight by overzealous overseers.

After many trials, a common solution to the problem of how best to organize the state's petroleum presence was achieved in all four countries. Government holdings gradually evolved toward a corporate institutional form. In all cases, these corporations have received a broad mandate. While Argentina and Brazil began with a direct ministerial dependency and Mexico and Venezuela started with more independent agencies, in all countries the state's petroleum interests passed through several phases before achieving a relatively stable institutional identity. Venezuela, with its comparatively late entry to the public petroleum enterprises, benefited from other countries' experiences and thus far has not attempted as many institutional variations as were tried elsewhere. While the motives for establishing a state petroleum presence varied and, once established, the agencies' objectives differed, the institutional outcome was similar—a dominant corporation, 100 percent government owned (Brazil excepted) but continually subject to internal reorganization.

Although the Latin American governments and their public petroleum enterprises have faced many similar problems, common treatment has more often been found at the level of company-government relations than in internal company practices. The emergence of company-specific management styles often led to views of problems, determination of strategy, application of tactics, and results characteristic of that company only. However, even in this diversity there is a common thread, as the degree of managerial autonomy has been severely limited and many of the internal management practices have been closely constrained by government-set parameters for action.

Organizational instability not only plagued the newly formed companies but recurred repeatedly throughout their history. This is in large part attributable to the lack of a public consensus about the role that public enterprises should play in an economy. Incoming regimes, cabinets, ministers, or boards of directors often imposed a new company structure, new procedures, and new managers. Ironically, from earliest times continual change became the only constant for most petroleum enterprises—change of company name, change of organizational goals, change of corporate form, change of management structure, and change in the company-government interface. In later years, everything but the name remained subject to change. While key individuals, such as Mosconi or Horta Barbosa, fought to maintain some institutional independence from external pressures to permit internal restructuring, after their relatively brief tenures outside interference increased. Only Petrobrás was able to maintain over time a relatively high degree of institutional identity and separateness. The other companies' successive transformations were all too often presented as responding to changes in security or nationalistic objectives and were rarely based in improving efficiency conditions.

The companies' organizational instability also derives from the muddled and varied objectives imposed on them over time. The common governmental roots of ignorance, uncertainty, self-interest, and fear of the future not only marked the formation of these petroleum companies, but they have also influenced every stage of their evolution. An unfortunate consequence of unreliability at the government level was that the externally imposed, continually modified targets for the public petroleum producers rarely were clearly defined, adequately stated, or readily attainable. Instead, these goals were often vague, mutually contradictory, contrary to management wishes, or politically motivated. Further, they were often imposed without regard either to other external constraints on company action or to company capacity to attain them. Changing goals usually led to changing at least part of the internal company structure to meet the revamped objectives. Even though maintaining a successful state oil industry required continually clear external policy direction, this was often lacking, a sad corollary to the lack of national consensus about the role of public enterprises. When strong individuals, such as Mosconi or Horta Barbosa, were able to set clear objectives that were externally ratified, they provided but a temporary respite from the normal organizational chaos.

Less-than-competent managerial decision making has often marked the public petroleum producers. There are several roots to this problem. First, the earliest major company figures were public enterprise entrepreneurs, more suitable to the rough-and-tumble environment of company organization and initial progress than to institutionalizing a strong management tradition. Second, accountability is strongly related to efficiency, and managers were not always held responsible for their decisions. Certainly with much important decision making removed from management discretion, as happened in Mexico and Argentina, there was little of substance for which they could have been held accountable. Third, a good incentive structure is strongly related to efficiency, but management was not always given incentives to improve performance. Again, the fault may be that performance indicators and standards were not always made explicit, while those that were explicit were frequently countermanded. Another contributing factor to an inadequate incentive structure was promotion based at times on politics, parentage, or patronage, not on performance. Fourth, competent managerial decision making is also inhibited by the presence of overbearing control systems. Although some public control must be exercised over public enterprises, governments generally have not recognized their companies' institutional prerogatives and, instead, have often usurped operational decision making from management. Lack of autonomy, lack of incentives—particularly pecuniary ones—and lack of consistent and attainable goals have often demoralized management.

Devising effective control systems for the public petroleum enterprises has proven to be a persistent source of difficulty. The public accountability approach to control seeks to correct management's deviations from national policy, while the company level approach seeks to promote internal efficiency. Although each can provide some increase in effective management for national objectives, both view the problem partially. Because public enterprises interrelate with multiple central government agencies and with many other public enterprises, key variables have been exogenous to the company or to its supervising ministry. For problems arising from interrelations with other government agencies, only a systems level approach could provide the basis for effective solutions, and that approach was nonexistent. While the presence of a key individual was at times useful in overriding obstructions from other agencies or in appealing directly to a higher (presidential) authority, this did little to institutionalize the mechanisms for dealing with systemic problems. This type of temporary solution only served to mask the problem.

Apportioning petroleum profits and economic rents has been a recurrent source of central government–public enterprise conflict. While governments generally adopted a strategy of trying to minimize the companies' retained earnings, confrontation usually arose when the companies tried to maximize control over reserves to expand operations and to maintain wages and salaries at internationally competitive levels. This perennial problem has generally been resolved in the government's favor by the explicit use of power, and the presence of a key figure at best mitigated the profit-sharing decisions taken elsewhere.

A consequence of this power-based cash transfer to central government has been persistently suboptimal petroleum investment levels. Inadequate capital transfers from governments marked the early company years, leaving exploration and exploitation underfunded. Later, governments for political reasons forced companies to adopt distorted pricing structures that did not adequately reflect the domestic scarcity of petroleum products. This produced markedly lower company income than would otherwise have been the case. Further, continued government appropriation of company surplus without regard to investment needs or to assigned objectives has been a constant factor in keeping investment low.

The current international anxiety about the debt crisis in Latin America has focussed attention on the debt position of public enterprises and, in particular, on that of the state-owned petroleum producers. Strong criticism from foreign and domestic sources has been levied at them for having borrowed abroad to finance investment projects or to pay for fuel imports. Although it is true that the public petroleum firms account for a large proportion of the foreign debt, to conclude from such evidence that public enterprise borrowing is out of control is a naive and simplistic error. Careful analysis shows that almost invariably the public enterprises were forced to borrow on international capital markets as a consequence of government interference in pricing and transfer mechanisms for political motives. As an example, it has been estimated that without price regulation Pemex 1979 sales would have risen to $20 billion, from $7 billion. That difference, for one year alone, far exceeded Pemex's total investments from 1971 to 1978.[41] In the past governments have been only too glad to force their public petroleum enterprises to borrow in the international money markets rather than to take the politically unpopular steps of raising prices or increasing taxes to the levels necessary to compensate the companies for the losses incurred in meeting their socially imposed goals. They further preferred siphoning funds from the companies to fund other government investment projects or even meet current expenses to covering such cash outlays by raising taxes, again often forcing the companies to borrow.

The previous discussion based on the analysis of four cases, including both importers and exporters, suggests that the Latin American public petroleum enterprises face problems that are not so much inherent in their government ownership structure but instead stem from the nature of the government-imposed environment within which they operate. While further detailed work is necessary to verify that these findings indeed apply to the other Latin American countries, there are indications that such is the case. Bolivia's YPFB, Chile's ENAP, Colombia's ECOPETROL, Ecuador's CEPE, Peru's PETROPERU, and Uruguay's ANCAP have all been forced to operate under adverse conditions imposed by their respective governments.[42] Although much less is generally known about the working environment of the small national oil companies in the Caribbean and Central America, the very absence of startling successes partly confirms my hypothesis.

If Latin America's public petroleum enterprises are to function more effectively within the future tighter integration to the world economy, several changes in their operating environments must occur. Government authorities outside the companies must clearly specify the company objectives, avoid constantly changing these objectives, and, if possible, carefully relate them to financial and other constraints. The authorities must examine the economic cost associated with each social goal and be willing to compensate the companies for meeting these goals. They must guarantee the companies sufficient political and economic autonomy to effectively insulate management from excess influence and interference. Finally, government authorities must institute mechanisms to adequately reward management for success or properly chastise it for failure.

Analogies to the public enterprises' multiple ambiguous roles, covered by Wirth in his introduction and underlying each of the four cases, hold true for governments. As mentioned earlier, the duality in governments' roles as owners of companies and owners of the petroleum reserves has given rise to perpetual conflict over profit sharing. A second polarity is between governments' roles as owners of companies and stewards of the nations' economic and political systems. Because public petroleum enterprises loom so large in a country's economy, because they provide employment for such a large portion of the labor force, and because their operations have such a significant political impact, the ownership dimension has often been neglected while the companies have been used to fulfill a variety of objectives for which they were not intended and remain unprepared to meet. A third duality is between governments as owners of companies and as managers of the companies they own. Governments have repeatedly demonstrated their inability to replace control mechansims more appropriate to sole proprietorships or small partnerships with the delegated authority structures so necessary in large modern corporations. This resembles private corporate boardrooms, where company founders are often unable and unwilling to relinquish the reins of control to their handpicked managers. Because governments have often refused to delegate control to public enterprise managers, the ownership dimension has been neglected while day-to-day government interference in management decisions has been commonplace. The lesson of the case histories of the Latin American petroleum companies' formative years is clear. The problems of the past have endured not because the public petroleum companies behaved *like the public enterprises they are;* they have endured because governments, *as public owners,* have abdicated their responsibilities to their companies.

Notes

The usual disclaimer applying, John Wirth's comments on an early draft of the conclusions are gratefully acknowledged.

1. "The Fortune Directory of the Largest Industrial Corporations outside the United States: The Foreign 500," Fortune 106, no. 4 (Aug. 23, 1982):182–201.

2. "The World's Largest Industrial Corporations," *Fortune* 108, no. 4 (Aug. 22, 1983):170–71.

3. "The Fortune Directory of the Largest Industrial Corporations outside the U.S.: International 500," ibid., pp. 170–89.

4. In the interest of streamlining this chapter and of not distracting the reader by excessive footnoting, only sources other than those cited in the earlier chapters will be included.

5. Although there is sketchy evidence from the analysis of non–Latin American cases that countries at similar levels of development have similar

public enterprise portfolios, the issue is conceptually different from the historical approach and will not be considered. See Leroy P. Jones and Edward S. Mason, "Role of Economic Factors in Determining the Size and Structure of the Public Enterprise Sector in Less-Developed Countries in Mixed Economies," in *Public Enterprise in Less-Developed Countries*, ed. Leroy P. Jones (New York, 1982), pp. 17–47.

6. E. V. K. FitzGerald, *The Public Sector in Latin America*, Cambridge University Centre of Latin American Studies Working Paper no. 28 (Cambridge, 1974); and Economic Commission for Latin America (ECLA), "Public Enterprises: Their Present Significance and Their Potential in Development," *Economic Bulletin for Latin America* 16 (1971), pp. 1–70.

7. George Philip, *Oil and Politics in Latin America: Nationalist Movements and State Companies* (New York and Cambridge, 1982), p. 318.

8. See, for example, the general taxonomies in Faqir Muhammad, "Public Enterprise and National Development in the 1980's: An Agenda for Research and Action" (paper presented at the Round Table on Public Enterprise of the Tenth General Assembly of the Latin American Association of Public Administration, Mexico City, 1978); Yair Aharoni, *Markets, Planning and Development: The Private and Public Sectors in Economic Development* (Cambridge, Mass., 1977); ECLA, "Public Enterprises"; and the specific dual categorization of Philip, *Oil and Politics*, p. 312.

9. Carl E. Solberg, *Oil and Nationalism in Argentina: A History* (Stanford, 1979), p. 16.

10. Philip, *Oil and Politics*, p. 163.

11. Rómulo Betancourt had advocated such measures as early as 1939, and before the 1948 military coups the probability of such action appeared high. See Edwin Lieuwen, *Petroleum in Venezuela: A History* (Berkeley, 1955), pp. 107–108.

12. Rómulo Betancourt, *Venezuela: Oil and Politics*, trans. Everett Bauman (Boston, 1979), pp. 394–95. The timing was important, however, as CVP was created at the same time that Venezuela was spearheading the preliminary negotiations that culminated in the formation of OPEC.

13. See, for example, Juan Pablo Pérez Alfonzo, *Petróleo y dependencia* (Caracas, 1971).

14. For a classification of government objectives sought through petroleum sector policies contrasted with multinational goals, including those of multinational public petroleum enterprises, see United Nations Centre on Transnational Corporations, *Alternative Arrangements for Petroleum Development: A Guide for Government Policy Makers and Negotiators* (New York, 1982), pp. 7–11.

15. Philip, *Oil and Politics*, p. 427.

16. Peter Seaborn Smith, *Oil and Politics in Modern Brazil* (Toronto, 1976), pp. 80–81.

17. Betancourt, Venezuela, pp. 394–95.

18. Philip, Oil and Politics, pp. 299–304.

19. International Bank for Reconstruction and Development/The World Bank, World Development Report, 1983 (New York, 1983), pp. 49–50.

20. For a comparative assessment of the profitability of large private and public enterprises, see Alfred H. Saulniers, "Public Enterprises in Latin America: The New Look" (paper presented at the seminar State Control and Planning of Public Enterprises, Brasilia, June 15–17, 1983).

21. Bryan Cooper, ed., Latin America and Caribbean Oil Report (London, 1979), p. 74.

22. A marginal tax rate of 50 percent means that the central government effectively pays half of all operating costs.

23. Cooper, Latin America and Caribbean Oil Report, p. 74.

24. Bruce Netschert, "Special Survey of Mexico's Oil and Gas Potential," in ibid., p. 138.

25. Cooper, Latin America and Caribbean Oil Report, p. 165.

26. Alfred H. Saulniers, "Public Enterprises in Mexico," Discovery 5, no. 4 (Summer 1981).

27. Philip, Oil and Politics, pp. 378–79.

28. Netschert, "Special Survey," p. 139.

29. United Nations, Alternative Arrangements, p. 25.

30. John D. Wirth, The Politics of Brazilian Development, 1930–1954 (Stanford, 1970), pp. 191f.

31. For examination of this hypothesis in nonpetroleum public enterprises, see Alfred H. Saulniers, "State Trading Organizations: A Bias Decision Model and Applications," World Development 9, no. 7 (July 1981); Saulniers, "ENCI: Peru's Bandied Monopolist," Journal of Interamerican Studies and World Affairs 22, no. 4 (Nov. 1980): 441–62.

32. Solberg, Oil and Nationalism, p. 64.

33. Philip, Oil and Politics, p. 168.

34. Ibid., p. 403.

35. Ibid., pp. 164, 403.

36. Ibid., p. 375.

37. Ibid., p. 164.

38. Ibid., p. 418.

39. Ibid., p. 375.

40. Ibid., p. 333.

41. The calculation, based on differentials between domestic and world market prices, takes no account of the income and substitution effects on consumer behavior. Nevertheless, it indicates that a substantially different pricing structure would have made much foreign borrowing unnecessary. See Saulniers, "Public Enterprises in Mexico," p. 22; Mexico, Secretaría de

Programación y Presupuesto, *La industria petrolera en México* (Mexico City, 1979).

42. See the respective chapters in Philip, *Oil and Politics*, and the chapter entitled "Government Involvement" in Cooper, *Latin America and Caribbean Oil Report*, pp. 71–88.

Appendix 1: Mexican Oil Industry Basic Statistics, 1910–80.
(million barrels/year)

	Production	Exports	Consumption	Refining	Imports
1910	3.6	0.7			
1911	12.5	0.9			
1912	16.5	8			
1913	25.7	21			
1914	26.2	23			
1915	32.9	25			
1916	40.6	27			
1917	55.2	46			
1918	63.8	52			
1919	87.1	76			
1920	157.1	146			
1921	193.4	172			
1922	182.3	181			
1923	149.6	136			
1924	139.7	130			
1925	115.5	99			
1926	90.4	81			
1927	64.1	53			
1928	50.1	33			
1929	44.7	27			
1930	39.5	27			
1931	33.0	23			
1932	32.8	23			
1933	34.0	22			
1934	38.2	25			
1935	40.2	25			
1936	41.0	25			
1937	46.1	15			
1938	38.8	4	28.2	33.7	1.7
1939	43.3	9	24.6	31.9	1.3
1940	44.4	9	28.0	32.1	2.2
1941	43.4	8	25.5	34.9	2.2
1942	35.1	1	27.0	33.2	2.4
1943	35.5	1	29.6	34.5	2.8
1944	38.5	1	31.3	36.5	3.3

(Continued)

Appendix 1: Continued

	Production	Exports	Consumption	Refining	Imports
1945	43.9	2	32.9	43.1	4.1
1946	49.5	3	38.2	51.3	4.5
1947	57.1	7	46.3	51.3	5.6
1948	59.8	7	45.7	49.8	5.3
1949	62.2	7	52.2	54.7	5.6
1950	73.9	12	58.0	55.9	7.3
1951	78.8	14	63.9	61.3	8.3
1952	78.9	9	65.2	65.1	7.7
1953	74.1	3	66.4	71.5	8.9
1954	85.2	5	68.2	77.1	11.5
1955	91.4	6	73.9	81.0	15.0
1956	94.1	7	80.8	80.8	18.0
1957	92.2	4	89.0	85.2	17.1
1958	100.6	1	94.4	94.3	12.4
1959	105.8	0.1	93.4	101.7	8.4
1960	108.8	1.1	101.4	102.4	7.9
1961	116.8	6.7	109.1	115.8	6.9
1962	121.6	7.1	105.3	116.8	6.0
1963	125.8	7.1	108.4	117.8	7.5
1964	129.5	7.6	117.5	127.0	9.6
1965	132.1	4.8	118.1	127.2	9.4
1966	135.0		121.1	129.5	12.8
1967	149.9		146.8	146.8	11.9
1968	160.5		143.9	155.9	11.8
1969	168.4		154.1	161.3	16.8
1970	177.6		162.7	175.6	17.3
1971	177.3		176.2	177.7	24.9
1972	185.0		200.5	193.4	25.5
1973	191.5		221.1	106.1	33.1
1974	238.3	6	240.4	234.3	23.6
1975	294.3	34	255.3	240.9	25.0
1976	327.3	34	272.9	268.3	15.7
1977	396.2	74	290.5	300.6	8.5
1978	485.3	194	320.8	320.2	13.6
1979	590.6	303	341.4	349.9	10.0
1980	709.0	401			
1981	844.0				

Sources:

1. Production. For 1910–79, Nacional Financiera, *50 años de revolución mexicana en cifras* (Mexico City, 1963), p. 68; Nacional Financiera, *La economía mexicana en cifras* (Mexico City, 1981), p. 64. For 1980–81, Petróleos Mexicanos, *Anuario estadístico 1981* (Mexico City, n.d.), pp. 37, 117f.

2. Exports. For 1910–38, G. Ortega, *La industria petrolera mexicana: sus antecedentes y su estado actual* (Mexico City, 1936), p. 34; R. Garcia Rangel, *El problema nacional petrolero* (Mexico City, 1939), pp. 115f. For 1939–80, Petróleos Mexicanos, *Anuario estadístico.*

3. Consumption. For 1938–79, Nacional Financiera, *La economía mexicana en cifras*, pp. 74–77.

4. Refining. For 1938–79, ibid., pp. 65–70.

5. Imports. For 1938–79, ibid., pp. 66–73.

Appendix 2: Venezuela Oil Industry Basic Statistics, 1920–83.
(thousand barrels/day)

	Production	Exports	Consumption	Refining	Imports
1920	1	2	1	1	
1921	4	3	1	1	
1922	6	5	1	1	
1923	12	9	1	2	
1924	25	23	1	2	
1925	55	50	1	3	
1926	98	90	1	4	
1927	166	156	1	7	
1928	290	275	1	14	
1929	373	358	1	13	
1930	371	367	1	14	
1931	320	307	1	17	
1932	319	302	1	18	
1933	324	310	1	20	
1934	373	357	1	22	
1935	407	380	1	25	
1936	423	411	1	23	
1937	509	460	1	24	
1938	515	489	5	26	
1939	560	518	6	37	
1940	512	429	7	73	
1941	621	611	8	87	
1942	416	386	8	63	
1943	409	477	8	61	
1944	702	682	11	72	
1945	886	870	12	89	
1946	1064	1082	16	96	
1947	1191	1161	25	101	
1948	1342	1282	29	119	
1949	1321	1260	40	145	
1950	1494	1424	52	250	
1951	1705	1612	57	315	
1952	1809	1715	62	348	
1953	1765	1662	75	413	
1954	1890	1789	90	443	

(Continued)

Appendix 2: Continued

	Production	Exports	Consumption	Refining	Imports
1955	2157	2024	106	537	
1956	2464	2312	123	624	
1957	2779	2576	142	689	
1958	2605	2437	131	732	
1959	2771	2578	134	824	
1960	2846	2685	128	882	
1961	2920	2764	122	923	
1962	3200	3018	151	1025	
1963	3248	3074	152	1042	
1964	3393	3213	166	1092	
1965	3473	3253	176	1175	
1966	3371	3182	180	1174	
1967	3542	3362	183	1221	
1968	3615	3369	194	1186	
1969	3594	3411	192	1156	
1970	3708	3470	201	1292	
1971	3549	3282	211	1245	
1972	3220	3065	228	1125	
1973	3366	3150	254	1303	
1974	2976	2752	249	1196	
1975	2346	2086	244	886	
1976	2294	2156	244	987	
1977	2238	1987	254	967	
1978	2166	1963	283	983	
1979	2356	2099	317	987	
1980	2168	1864	355	922	
1981	2107	1759	369	859	
1982	1893	1554	381	866	

Source: Official figures supplied to Edwin Lieuwen by Petróleos de
Venezuela.

Appendix 3: Argentine Oil Industry Basic Statistics, 1910–78. (million barrels/year)

	Production	Exports	Consumption	Refinery Capacity	Imports
1910	0		1.0		1.0
1911	0		1.1		1.1
1912	0		1.3		1.3
1913	0.1		1.9		1.8
1914	0.3		1.8		1.5
1915	0.5		3.4		2.9
1916	0.9		3.2		2.3
1917	1.2		3.4		2.2
1918	1.4		2.5		1.2
1919	1.3		4.7		3.4
1920	1.7		6.7		5.0
1921	2.1		8.2		6.1
1922	2.9		9.4		6.6
1923	3.3		10.8		7.5
1924	4.7		12.8		8.1
1925	6.0		11.4		5.3
1926	7.9		14.8		6.9
1927	8.6		17.5		8.8
1928	9.1		19.8		10.7
1929	9.4		21.4		12.0
1930	9.0		21.6		12.6
1931	11.7		21.6		9.9
1932	13.2		20.2		7.1
1933	13.7		20.4		6.7
1934	14.0		21.9		7.8
1935	14.3		24.1		9.8
1936	15.5		25.6		10.1
1937	16.4		28.1		11.7
1938	17.1		31.9		14.8
1939	18.6		32.8		14.2
1940	20.6		33.7	37.6	13.1
1941	21.7		35.3		13.6
1942	23.7		31.1		7.4
1943	24.9				
1944	24.3				

Appendix 3: Continued

	Production	Exports	Consumption	Refinery Capacity	Imports
1945	22.9				
1946	20.8				
1947	21.9				
1948	23.3				
1949	22.6				
1950	23.5		56.7	55.5	33.2
1951	24.5				
1952	24.9				
1953	28.5				
1954	29.7				
1955	30.6		70.1		39.6
1956	31.1		71.1		40.0
1957	34.0		75.5		41.5
1958	35.7		80.5		44.8
1959	44.6		76.1		31.4
1960	64.0		90.3	86.9	26.3
1961	84.6		110.4		25.8
1962	98.4		117.9		19.5
1963	97.3		108.0		10.7
1964	100.4		114.3		13.9
1965	98.4		130.6		32.1
1966	104.9		139.0		34.0
1967	114.9		137.2		22.4
1968	125.7		139.4		13.7
1969	130.1		145.3		15.0
1970	143.6		164.2	183.2	20.6
1971	154.8		173.7		19.0
1972	158.7		170.0	229.9	11.3
1973	153.7		175.1	219.7	21.4
1974	150.7		172.2	227.8	21.6
1975	144.5		160.1	263.2	15.6
1976	145.6		167.8	249.7	22.2
1977	157.5		179.0	256.6	21.5
1978	165.2		180.0	239.1	15.6
1979				239.1	
1980				246.7	

Sources:

1. Production. For 1910–37, "Resumen estadístico de la economía
argentina," *Revista de Economía Argentina* 20 (Nov. 1938): 323. For
1938–42, Argentina, Comité Nacional de Geografía, *Anuario geográfico
argentino, suplemento 1942* (Buenos Aires, 1943), pp. 123–24. For 1943–
66, Instituto Argentino del Petróleo, *El petróleo en la República
Argentina* (Buenos Aires, 1967), p. 5. For 1973–78, Carlos Villar Araujo,
"Informe sobre el petróleo en la Argentina," *Crisis* 24 (April 1975): 14.
For 1973–78, Frank E. Niering Jr., "Private Capital Aids Oil Drive,"
Petroleum Economist 46 (Oct. 1979): 409.

2. Imports. For 1910–37, Adolfo Dorfman, *Evolución industrial argentina*
(Buenos Aires, 1942), p. 143. For 1938–42, *Anuario geográfico,
suplemento 1942*, p. 124. For 1950 and 1955, *International Petroleum
Encyclopedia* (Tulsa, 1972), pp. 240, 292. For 1960–66, James E. Zinser,
"Alternative Means of Satisfying Argentine Petroleum Requirements," in
Foreign Investment in the Petroleum and Mineral Industries, Raymond F.
Mikesell, et al., (Baltimore, 1971), p. 191. For 1967–72, *International
Petroleum Encyclopedia*, pp. 240, 292. For 1973–78, Niering, "Private
Capital," p. 409.

3. Consumption. In all cases this equals production plus imports.

4. Refining Capacity. *International Petroleum Encyclopedia*, p. 301.

Appendix 4: Brazilian Oil Industry Basic Statistics, 1920–80. (thousand barrels/day)

	Production	Exports	Consumption	Refining Capacity	Imports
1920			6.6		6.6
1921			7.8		7.8
1922			5.9		5.9
1923			6.6		6.6
1924			8.9		8.9
1925			10.7		10.7
1926			10.0		10.0
1927			14		14
1928			12		12
1929			16		16
1930			15		15
1931			14		14
1932			12		12
1933			15		15
1934			16		16
1935			17		17
1936			19		19
1937			22		22
1938			23		23
1939			25	4	25
1940			25		25
1941			21		21
1942	0.1		15.1		15
1943	0.1		15.1		15
1944	0.2		15.2		15
1945	0.2		19.2		19
1946	0.2		32.2		32
1947	0.3		48.3		48
1948	0.4		61.4		61
1949	0.3		69.3		69
1950	1		85	12	84
1951	2		104		102
1952	2		121		119
1953	3		130		127
1954	3		154		151

(Continued)

Appendix 4: Continued

	Production	Exports	Consumption	Refining Capacity	Imports
1955	7		178	106	171
1956	11		197		186
1957	28		198		170
1958	52		247		195
1959	65		250		185
1960	81		270	208	194
1961	95		280		210
1962	91		310		223
1963	98		340		231
1964	91		350		235
1965	94		330	364	221
1966	116		380		243
1967	147	1.6	380	380	226
1968	164	0.3	460		286
1969	175	0.5	480	501	281
1970	165	17	510	500	329
1971	170	30	560	565	398
1972	165	42	650	795	474
1973	165	71	770	795	663
1974	175	40	830	1020	681
1975	170	38	870	1020	706
1976	170	56	970	1020	827
1977	165	40	1010	1175	827
1978	165	28	1050	1230	910
1979	170		1180	1230	—
1980	185		—	1375	999
1981	213		—		812
1982	268		1020		696
1983			950		

Sources: Instituto Brasileiro de Geografia e Estatística, Anuário estatístico do Brasil, 1960 (Rio de Janeiro, [1960]), p. 50, and later issues. Crude Production and refinery capacity (both 1970–80) are from British Petroleum, BP Statistical Review of the World Oil Industry, 1980 (London, n.d.). Consumption Data for 1960–80 are from U.S. Department of Energy, Energy Information Administration, 1982 Annual Energy Review (Washington, April 1983), 83. Production data for 1981–82 and import data for 1980–82 are from EIU, Quarterly Energy Review, Latin America and the Caribbean 1 (London, 1983): 32. 1982–83 consumption data are from EIU, Quarterly Economic Review of Brazil 2 (London, 1983): 14.

The Contributors

Jonathan C. Brown is associate professor of history at the University of Texas, Austin.

Esperanzo Durán is a research professor at the Centro de Estudios Internacionales in El Colegio de México and is currently a Research Fellow at the Royal Institute of International Affairs.

Edwin Lieuwen is professor of history at the University of New Mexico.

Alfred H. Saulniers is Coordinator, Office for Public Sector Studies, the Institute of Latin American Studies, University of Texas at Austin.

Carl E. Solberg, who died in April 1985, was professor of history at the University of Washington, Seattle.

John D. Wirth is professor of history and former director of Latin American Studies at Stanford University.

Index

www.ingramcontent.com/pod-product-compliance
Lightning Source LLC
Chambersburg PA
CBHW021550210326
41599CB00010B/381